SURFACE ENGINEERING

HOW TO ORDER THIS BOOK

BY PHONE: 800-233-9936 or 717-291-5609, 8AM–5PM Eastern Time

BY FAX: 717-295-4538

BY MAIL: Order Department
Technomic Publishing Company, Inc.
851 New Holland Avenue, Box 3535
Lancaster, PA 17604, U.S.A.

BY CREDIT CARD: American Express, VISA, MasterCard

SURFACE ENGINEERING

PROCESSES AND APPLICATIONS

Edited by

Ken N. Strafford
Roger St. C. Smart
Ian Sare
Chinnia Subramanian

TECHNOMIC
PUBLISHING CO., INC.
LANCASTER · BASEL

Surface Engineering
a **TECHNOMIC**®publication

Published in the Western Hemisphere by
Technomic Publishing Company, Inc.
851 New Holland Avenue, Box 3535
Lancaster, Pennsylvania 17604 U.S.A.

Distributed in the Rest of the World by
Technomic Publishing AG
Missionsstrasse 44
CH-4055 Basel, Switzerland

Printed in the United States of America
10 9 8 7 6 5 4 3 2 1

Main entry under title:
   Surface Engineering: Processes and Applications

A Technomic Publishing Company book
Bibliography: p.
Includes index p. 347

Library of Congress Catalog Card No. 94-61268
ISBN No. 1-56676-154-9

Table of Contents

Preface

THIS book has its origins in an international conference on surface engineering held in Adelaide, South Australia, in 1991. The conference brought together specialists in a variety of surface engineering techniques and applications, together with industrialists, reviewing the current implementation of many of these processes. International experts in the methods and techniques for characterisation of surface coatings and modified surfaces also presented reviews of the status of these performance indicators. In addition to descriptions and appraisals of the theoretical and experimental principles applying to surface engineering processes, the contributors also discussed limitations to the uses and applications of the various techniques in materials technology and engineering.

The book, thus, provides updates on the many processes and applications reviewed at the conference. It contains a selection of the plenary papers and a variety of particular examples of surface processes relating to wear, corrosion, and other surface modifications.

As such, the intended readership of the book comprises four major groups:

- senior undergraduate students in materials engineering, metallurgy, materials science, and surface chemistry and physics
- postgraduate students undertaking research in surface-engineered materials, surface analysis, and related materials science
- groups of scientists and engineers in research and development seeking a description of accessible surface engineering techniques and guidance on the most appropriate methods for their particular applications
- industrial scientists and engineers requiring updating or

undertaking training courses and workshops on new materials engineering involving surface treatments

The contributors to the book comprise a cross section of Australian and international experts. They are all closely involved with the international engineering and scientific communities in their own areas of research, and this is reflected in the selection of examples in each chapter. Chapters emphasize that a very wide spectrum of methods and techniques for surface engineering is now available. One aim of the volume has been to encourage people involved in research, development, and process control to become aware of the increasing availability and utilization of these treatments in their own fields of materials science and engineering. Equally, the wide availability of testing methods, new techniques, and instrumentation is illustrated in Part II for the assessment of existing and newer processes. It is anticipated that this book will also serve a need by extending the awareness of the range of processes, applications, and characterisation methods from those covered by more specialized texts confined to particular processes and topics separately and, at the same time, provide examples of real applications to research and development in surface engineering implementation.

KEN STRAFFORD
ROGER ST. C. SMART
IAN SARE
CHINNIA SUBRAMANIAN

Introduction

DEFINITION AND SCOPE

SURFACE engineering, now recognised as an enabling technology of major importance in the successful, most effective, and efficient exploitation of materials in engineering practice, may be defined in a number of ways. However, broadly, the subject concerns at least three separate and yet interrelated activities—*the optimisation of surface properties* (concerned particularly with the performance of surfaces and coatings in respect to corrosion and wear); *coatings' and modified surfaces' characterisation* (with respect to condition; composition; structure and morphology; and mechanical, electrical, and optical properties); and *coatings' and surfacing technologies* (embracing the more traditional technologies of painting, electroplating, weld surfacing, and spraying, as well as thermal and thermochemical treatments such as nitriding and carburising (especially in their plasma variants)), and the more recently emergent technologies such as laser surfacing, physical and chemical vapour deposition, and ion implantation. Noteworthy in many of these technologies has been the recognition of the considerable advantages in the adoption and use of plasma-assisted processing.

It is evident from the mass of literature that has emerged in recent years, and especially over the last decade or so, that the scope and potential of surface engineering is enormous and in some ways perhaps limited by the power of the imagination rather than technical difficulties. At the same time, as in many other technologies, the lack of understanding at the basic level, particularly in the overall design and serviceability of surface engineered artifacts, both in terms of the advanced processes (e.g., the physics of the ion plating processes) and the structure/property relationships of/within the surface treatment/coating (and associated substrate, i.e., the system) is a major handicap in optimisation—*fitness-for-purpose*. It is now possible to deposit almost any material on an appropriate substrate (which itself may, in addition, be "engineered," e.g., hardened to create a duplex surface engineered component) to produce a coating system for optimum performance under very specific working conditions. This flexibility of modern surface engineering technology poses, however, real problems because of the almost infinite combinations possible, and there is clearly a need to develop a much more systematic design methodology—expert systems for coating and surface treatment selection. There is also a related need for effort in education and training to make engineers much more aware of the potential of surface engineering in developing resistance to corrosion and wear processes recognising that failures commonly are initiated at ill-defined, "unsuitable" surfaces.

SURFACE ENGINEERING: A RAPIDLY DEVELOPING ENABLING TECHNOLOGY

The term *surface engineering* (SE) is of relatively recent origin and use. However, the use of coatings and surface treatments to develop specific "tailor-made" surfaces to respond most effectively and efficiently to engender satisfactory design life serviceability in engineering artifacts is not new. Indeed, it is possible to recognise a whole historical perspective to surface engineering practices, such as the ancient pre-Christ art of carburising to yield sharp, enduring, cutting edges in weapons and tools, the major developments in the early and mid-nineteenth century in Europe in processes such as electroplating and galvanising, to the very recent drive and impetus in electronic materials coatings over the past twenty five years or so, so crucial to semi-conductor device technology, which, via the associated chemical (CVD) and physical vapour deposition (PVD) processes, has, more than any other single technology, allowed the advent and development of advanced hard metal tribological coatings, perhaps epitomised by titanium nitride.

PVD processing, in its many variants, through its very rapid advance over the last decade or so, provides, inter alia, the potential, because of its extreme flexibility as to coating type, for a major breakthrough in the design of novel coating systems for specific requirements. This is undoubtedly of major significance, recognising the specificity of coatings to overcome particular wear or corrosion problem situations. Of special interest here are the exciting possibilities for multilayer coating systems providing for the matching of the normally totally different requirements of coatings at the working interface/environment and the substrate/coating interface.

In all these situations, however, it is important to recognise the need for a sufficient R&D effort in the processing technology itself—thus, in general, there is poor understanding of the mechanisms of evaporation, ionisation, and deposition at the fundamental level in PVD processing. Equally, the level of comprehension of the basic relationships and linkages between coating properties and characteristics, and performance and serviceability is low (structure/property relationships) in contrast to the systematic and detailed knowledge now generally available for similar materials in bulk monolithic form. Nevertheless, it is evident that SE, via the use of its many existing and developing variants, provides real opportunities to "tailor make" surfaces (involving recognition, ab initio, at the design stage) for optimum, economic performance and service.

TECHNICAL MANAGEMENT AND THE BUSINESS OF SURFACE ENGINEERING

Doyle [1] has provided a valuable insight into managing the business of surface engineering, offering a processing and supplier perspective of the world of surface engineering. Clearly, the range of activities in surface engineering is many and varied, which in turn, demands many and varied skills on the part of the manager. Much of the surface engineering business is done on a subcontract basis and it is, therefore, a service industry. In this industry, it is the customer's needs that must be met. This demands a total customer responsiveness in which the business relationship is more of a partnership than one of adversarial dealings. In this partnership, there are many factors that must come together to make it a mutually successful business. Doyle [1] has identified three basic building blocks that go to making a solid foundation to this business partnership, namely (1) the recognition of the need to provide the best quality, (2) the need to provide an expert-based knowledge to the customer, and (3) the requirement to be more aware of the technical innovations and developments in their own and related technologies.

Some of the issues concerning QA/QC in SE are also raised below within the context of the users' dimension and perception of quality. Doyle points out that the surface engineer faces a challenging time in addressing the issue of quality. Indeed, for some, this is an even more daunting task—for example, how to provide a quality assessment of a 2 μm thick PVD coating or a few nanometer thick ion implanted surface layer. The challenge is one of providing the customer with a measure of certain parameters that will meaningfully relate to the function and performance of the surface coating.

PVD coated tools have a built-in quality that derives from the operational method of coating. The PVD process is carried out in vacuum, and, therefore, all tools are rigorously inspected for contaminants. At this stage, tools with defects such as burrs or grinding damage are rejected. Tools accepted for coating are then cleaned thoroughly, coated and a further inspection follows. The PVD process is a batch processing operation with well-defined system parameters, such as pressure, current, gas flow, etc., which are all recorded. Overall, PVD coated tools are therefore of very high quality, and it remains only to establish procedures that unambiguously deliver a quantitative measure of, e.g., adhesion.

The surface engineer is often the last person in the line of manufacture to put the finishing touches to what is a complex and costly piece of tooling. In this transaction, the customer is placing great trust in the surface engineer to do the job properly. It is essential that the manager instill in his staff a total quality consciousness.

Doyle [1] notes that an important element in the business relationship between the surface engineer and the customer is the exchange of *expert-based knowledge*. Primarily, this must relate to the service or services being offered but must also cover some of the broader issues—for example, material engineering treatments depend on the correct choice of the substrate and its thermal history. Frequently, if the facts are presented simply, without the smokescreen of ignorance, then there is a synergistic effect resulting from the customer's own expert knowledge of his products or tooling that often leads to further improvements.

Good communication is perhaps even more important when dealing with some of the newer technologies involving ion beams or plasmas. These treatments demand a level of understanding of physics and chemistry only briefly touched upon in most engineering education and courses. The importance of education and training courses in surface engineering is discussed below. In this situation, an atmosphere of trust and mutual respect is essential, particularly when the surface engineer may be introducing novel technology as a solution looking for a problem. The surface engineer is then relying on good feedback in order that a business be established.

Finally, Doyle [1] has emphasized how, in the management and operation of the business of SE, due attention must be paid to company personnel being aware of technological development and innovation in the field. Here he suggests there are two aspects. Firstly, the manager should be aware of innovations and developments that directly impinge on markets and, secondly, the manager should take a more forward-thinking attitude in appreciating the importance of research and development to ensure the company's long-term survival.

He believes that, with respect to PVD technology, it is particularly important to be able to point out to the customer the commercial realities from the whole sea of technological possibilities. A few years ago, there was much speculation that titanium aluminium nitride would generally displace titanium nitride. This has not so far eventuated for several reasons, not the least of which has been the technical difficulty of producing a ternary coating by PVD means. A ternary coating of titanium carbonitride will, however, become increasingly available to industry over the next few years. Its higher hardness than titanium nitride may make it more suitable for cutting tools used on the highly abrasive aluminium alloys.

Doyle [1] also noted, in the context of technological development and innovation, one area of potential development for PVD coatings is in the area of corrosion resistance. Unfortunately, PVD coatings contain process-generated defects that degrade the integrity of the coatings. Thus PVD titanium nitride coatings perform poorly in salt spray tests. Research on arc evaporation, for example, by CSIRO in Australia, offers the possibility of eliminating one of the major process defects, i.e., macroparticles. A combination of this development with a more noble metal in the titanium melt, such as palladium, could add a further dimension to the use of PVD coatings. These developments are, however, some time away from commercial exploitation.

EDUCATION AND TRAINING

Despite the fact that surface engineering is of critical importance to the manufacturing and fabrication industries and application areas such as mining, power generation and distribution, and automotive and aerospace, specialist technicians or graduate engineers in surface engineering are not yet produced or recognised. In fairness, however, it should be pointed out that various specialist courses in a number of subject areas subsumed within surface engineering—such as corrosion engineering and tribology—especially, at the postgraduate level, have been in existence for a number of years in several universities worldwide. Again, the importance of the field of advanced manufacturing technology, within the context of materials processing, production, and associated tooling, has also been well catered for. Finally, the growth of the subject of materials engineering over the past fifteen to twenty years should also be recalled, with its built-in emphasis on design and materials selection wherein, it would be anticipated, there would be appropriate attention to the tole of surface coatings and treatments within the total design audit. Indeed, Strafford et al. [2] have argued that the SE dimension in design must be embraced ab initio in order to be able to optimise component design—"fitness for purpose."

Just, then, what are the needs for education and training in surface engineering? At the outset, it has to be remembered that SE is a blend of applied science and engineering and is interdisciplinary in nature. Thus, the subject, in some of its most advanced forms, such as PVD processing, requires an understanding, or at least an awareness, of the realms of plasma physics, the physical and chemical properties of materials, their mechanical properties, and especially structure/property relationships in real materials. It has to be recognised, however, that while the world of materials science and physical metallurgy has benefitted over the past twenty-five to thirty years (so that is is now possible to design materials for specific functioning), much less is known of the basic science and properties of coatings, especially the thin coatings such as the nitrides, carbides, and oxides of potential interest in tribological situations. Hence, the educator and trainer are at a disadvantage in detail, albeit a basic rationale and methodology exist: on the other hand, the student is fortunate, perhaps, to be involved in the forefront of knowledge, with its excitement of discovery. This is particularly true of the research student.

Ultimately, the coatings' user is concerned with enhanced performance, most commonly in the areas of tribology and wear and corrosion/environmental behaviour. It follows that students must become totally familiar with these fields.

In order to provide focus and appropriateness as to detailed areas of study in these large and complex fields, it is perhaps most useful to approach and enter the subject via an analysis of the detail of the failure mechanisms occurring through wear and corrosion, etc.—a forensic analysis—in order to be able to identify key aspects of necessary background knowledge—particular weaknesses in design or serviceability of engineering components becoming known and categorized and recognised through experience.

There is plainly a need to educate and train scientists and engineers at various levels—technician, graduate, and postgraduate. Because of the multidisciplined nature of surface engineering, it is clearly of benefit to design courses to accommodate persons of widely differing backgrounds—such diversity in a class or student intake should not be seen as a source of difficulty in a simple approach to teaching (as, in actual fact, it will be), but rather a challenge, mirroring the breadth of the subject. This diversity in intake is believed to be of major potential benefit, in fact, in postgraduate courses or research where a blend of student age, background, and experience can be a major plus, for example, in integrating the approach to a design project. In real-life situations, in the solution of tribology or corrosion problems, the ability and knowledge of the specialist to synthesize and move easily in relevant areas of the constituent subject elements is mandatory, and must be a principal objective in any education and training course in SE.

THE RESEARCH PERSPECTIVE IN SURFACE COATINGS SCIENCE AND TECHNOLOGY

SCOPE AND PURPOSE

Surface coatings have been employed for many years to modify locally the base properties — chemical, physical, or mechanical — of an artifact so that its working performance and serviceability may be acceptable in relation to design life requirements, usually in terms of tribological and/or corrosion performance. Thus, in historical context, as early as 1800 one Professor Brugnatelli at Davia University described silverplating, following on from Volta's basic studies on the production of electricity by metallic corrosion in 1786 [3]. By 1840 commercial silver electroplating was well established. Electroplating of nickel using the Watt's bath dates from 1916, and this process is probably the best known technique used to provide a basis for the protection of mild steel against corrosion in a great variety of engineering situations.

Nickel-based coatings may also be used to enhance wear resistance, as well as providing corrosion resistance. In these situations, however, electroless nickel coatings are employed [4–7]: this technology is of more recent origin (1946) and remains a subject of intense interest [8] and wide applicability.

The potential offered by the use of composite electroless coatings is becoming increasingly recognised, offering an additional dimension in the design of coating — for example, the incorporation of PTFE particles in electroless nickel engenders lubricity and reduces friction [9–11], while admixture with SiC granules enhances abrasion resistance. Clearly, there is the prospect of satisfying, simultaneously, requirements for both corrosion and wear resistance by suitable choice of coatings and attention to composition microstructure, heat treatment condition, etc. Optimisation — fitness for purpose — will only be achieved by a better understanding of factors determining such properties at the fundamental level.

Again, it is fascinating to note historical context that, while the technology of zinc coatings, in its many process variants, including electroplating, has been in use for many years for the corrosion protection of steel, it is also an area still undergoing rapid sophisticated developments — for example, in the development of Zn-Co, Zn-Ni, or Zn-Fe coating alloys for improved corrosion protection, permitting the use of thinner fabricable and weldable sheets of particular interest in the automotive industry in the drive towards lighter, more energy efficient vehicles. It is believed here that additions of $\sim$ 10–15% Co, Ni, or Fe encourage passivity [12], a phenomenon requiring detailed study and clarification.

Coatings technology is merely one, albeit major, weapon in the engineer's armoury of surface engineering techniques, a subject that also includes the techniques of surface finishing (such as anodising) and heat treatment. Strafford and coworkers [13] have suggested that the term *surface engineering* encompasses all of these techniques and processes, which are utilised to induce, modify, and enhance the performance—such as wear, fatigue, corrosion, resistance, and biocompatibility—of surfaces covering the three major inter-related activities of (1) the optimisation of surface properties for specific enhancement of performance, (2) coatings' process technologies, and (3) coatings' characterisation. Activity (1) is, in reality, also concerned with the performance of the substrate, i.e., the coatings system as a whole, in terms of corrosion, adhesion, wear, and other physical and mechanical properties. Beyond the more traditional (albeit still developing) technologies of painting, elctroplating, weld surfacing, plasma and hypervelocity spraying, and various thermal and thermochemical treatments, within (2) are the newer technologies of laser surfacing, physical and chemical vapour deposition, and ion implantation.

Activity (3) also concerns the evaluation of surfaces and interfaces in terms of composition; morphology and structure; and mechanical, electrical, and optical properties. It is concerned ultimately with a fundamental understanding of structure/property relationships in coatings materials and their optimum design—fitness for purpose—via microstructural engineering.

While there is a natural intrinsic interest in understanding fundamental structure/property relationships in coatings—for example, in the mechanisms of the hardening of electroless high boron nickel-boron amorphous coatings for wear resistance, brought about by heat treatment, or in the potential again of the amorphous coatings of alloys, which may be produced by pulse-electroplating, offering greatly improved corrosion resistance—more pragmatically there is the drive from the commercial and industrial world where added value, either directly or indirectly, through the improved serviceability and reliability of coated components, is the prize.

In this context, it has been suggested [14] that in the UK, for example, a market for £570M/year existed in 1989 for process technologies, both established electroplating (welding, nitriding/carburising, and metal spraying) and emergent (ion implantation, plasma chemical and physical vapour deposition, plasma spraying, laser/electron beam processing, and friction surfacing). It was believed that the market value of UK products, which were critically dependent on surface engineering as an enabling technology, could be as high as £5–10B per annum. In addition, a major source of concern in the UK was the very low level of emergent process exploitation, estimated at less than 2% of direct market value. This was in contrast to uptake in Germany, France, Japan, and the USA. A recent Australian survey conducted by Wilks and Strafford [15] has also highlighted the overall importance of surface engineering technology in the manufacturing economy, but also, as with the UK, the present low level of knowledge, expertise, and take-up of the advanced emerging SE process technologies.

An additional illustration of the importance of surface engineering practices in financial terms is the estimated cost of corrosion to the UK economy of some £6,500M per year. Again, it is believed that proper attention to tribological problems could lead to savings of between 1.3 and 1.6% of the UK Gross National Product. The enormous benefits and competitive advantages associated with the use of sophisticated coated tools, where the total design of the tool itself and its associated coatings, possibly in multilayer format, that is, the coating system, have been widely discussed—see for example [16]. Here the potential for enhanced quality and surface finish are noteworthy.

THE RESEARCH DIMENSION: AN OVERVIEW

Strafford and coworkers [13] have provided a conspectus of surface engineering—a broad term defined earlier, but one where the significance, contribution, and role of surface coatings was emphasized. In particular, they have enunciated and discussed the *research dimension* pertaining to the subject. It has been pointed out that, as with many rapidly developing areas of technology, the understanding of the underlying basic science lags behind what are often spectacular practical improvements in properties achievable through surface engineering. While it may be argued that much can be, and has been, achieved in the absence of a clear research rationale for surface engineering over the years, embracing, of necessity, many areas of the physics and chemistry of materials, the degree of complexity increasingly required for coatings certainly compels the establishment of a theoretical framework. Research is needed, not only to characterise and understand the functional behaviour of coatings, but also at a more fundamental level involving the systematic application and testing of scientific principles, with a view to creating entirely new types of surfaces with novel properties. Without more attention to research and development, a full realization of the potential of this technology cannot be achieved. More pragmatically, a research and, especially, development effort is also required in the achievement of more reliable coatings, with reproducible properties and in the economic necessity for appropriate levels of QA and QC.

In the evolutionary path to enhanced properties by surface engineering processes, a number of interconnecting functions may be distinguished [13]. In the first place, the microstructure and composition of the engineered surface coating and the substrate (prepared with respect to standardised, as-established levels of stress and contamination, structure, composition, and topography) are optimised to meet particular engineering demands. Historically, single-layer films have been deposited; however, increasing levels of competence, confidence, and sophistication, especially in process technology, have earmarked the development and hold the real prospect of creating complex, multilayer (hybrid) films, capable of satisfying a combination of discrete properties requirements that constitute the component's operating envelope.

Central to the full utilization of such multiform, multicompound treatments is the full development and efficacy of technological processes, be they CVD- or PVD-based, with supplementary plasma, laser, or ion-assistance. Here, understanding, characterisation, and control of surfaces and, especially, their reproducibility and reliability to acceptable standards of quality control and assurance, are crucial in the development of new coatings systems from laboratory to full-scale commercialization.

The properties of the surface, substrate, and interface are critically related to several often mutually dependent parameters, e.g., characteristics of the operating process and sample topography. Added value to the component in relation to enhanced mechanical, thermal, chemical electrical, or optoelectrical attributes, as well as fitness-for-purpose and the minimization of life cycle costs, are the driving forces for the justification, rationale, and adoption of surface engineering.

The ability to predict the performance (e.g., fracture mechanisms and time-dependent behaviour) of surface modified artifacts using surface analytical techniques, system modelling, interfacial simulation, and NDE (i.e., characterisation) is essential if the engineer and designer are to fully exploit the potentials of the discipline. The first iterative cycle ends with the up-scaling (i.e., pilot programmes) from laboratory-sized developments. Design methodologies, possibly augmented with computerized data bases/expert systems, facilitate (timely) refinement.

A RESEARCH PROGRAMME: RATIONALE, SCOPE AND CONTENT

It is evident from the foregoing that a major research effort in the area of advanced surface coatings is timely and necessary in order for maximum benefit to be realised, especially in competitive manufacturing industry. Advanced materials have long been recognised as a major factor in determining industrial growth worldwide [16]; it is evident that an essential and crucially important element within the broad subject concerns the science and technology of surface coatings. The successful application of a suitable surface coating, tailored to meet specific additional demanding requirements at an engineering artifact's interface with its working environment, often will determine the ultimate practicality of a particular design – certainly it will determine the artifact's serviceability and reliability. It has been pointed out by Strafford et al. [11], that, while the present intense and growing interest in surface engineering may be seen as only part of a continuous spectrum of activity spanning thousands of years, the prospect of true "fitness-for-purpose" in many engineering scenarios will only be achieved by a much fuller basic understanding of structure/property relationships in coatings and their interaction with the substrates (the coating system). Much of the systematic fundamental approach in the research methodology embracing the subject of materials science, especially over the past twenty-five years or so, may be adopted, with appropriate adaption to the world of surface coatings. Thus, the principles of alloy design, for example, now established on the basis of this research effort in physical metallurgy, need to be applied to coatings' design and processing formulation.

Strafford et al. [17], for example, have discussed the design of alloys to resist high-temperature environmental degradation processes – oxidation, sulphidation, and chloridation – all processes that presently limit proposed advances in more efficient fossil fuel energy conversion devices. In particular, they have enunciated, on the basis of many years' published research effort, guidelines as to alloy chemistry for optimisation of performance, taking into account the subtleties of the often novel and usual working environments pertaining to several of such devices, e.g., gasifiers. The potential of such novel alloys for service as high-temperature protective coatings has also been noted where the limiting problem to ultimate success is likely to be interdiffusion between coating and substrate: attention thus needs to be focussed on the design of diffusion barrier materials – perhaps based on the Group V, VI refractory metals. In this context, there is a dearth of fundamental information – diffusion data – and basic directed studies are required before novel barrier layer/coatings can be confidently identified. Physical vapour deposition process technology has attraction in being able to produce, in principle, the necessary multilayer coating systems.

This area – novel high-temperature coatings systems – is merely one important example of a key research area involving a balanced approach to the relevant science and technology, following this iterative interaction [2]. Other important timely areas, broadly relating to the control of wear and corrosion are expected to benefit from such balanced research and development activities. Here the industrial input is crucial: the research programme is about solving real industrial problems, drawing in and researching the appropriate fundamental aspects of materials science and translating this through modern advanced (process) technologies (such as PVD) into working practical solutions in collaboration with industry.

SUMMARY

In summary, considered main objectives of a comprehensive research effort in surface engineering may be cited as follows:

- to systematically review and analyse fundamental aspects of certain tribological and corrosion phenomena (interfacial compatibility problems) in identified selected critical areas occurring in the manufacture and serviceability of various engineering artifacts
- to examine, ab initio, theoretical aspects of the design of advanced novel surface coatings and coatings systems, and so to develop a rationale for the inhibition of such phenomena, using coatings, based on appraisal of key coatings materials characteristics both natural and engineered (structure/property relationships) and the physical and chemical nature of the various aggressive/ incompatible working environments and counterfaces
- to develop an optimised realistic and practical approach to this design process, taking into account known and especially emerging advanced coating process technology capabilities
- to conduct a structured and focussed experimental research and development surface coatings programme – preparation (embracing a number of different advanced process technologies), characterisation, and performance evaluation in selected identified key areas of current and emerging interest and concern, of direct relevance to the world of industry and manufacture – coatings' equipment manufacturers; suppliers and producers of coatings; coatings' users in manufacturing processes; and end-user enterprises concerned with improved life, reliability, and serviceability of coated componentry
- through this experimental programme to draw up and recommend quality assurance parameters for advanced surface coatings and coatings' systems, based upon systematic experimental assessment of the linkages between identified significant coatings' properties and characteristics, and performance – in the laboratory and in actual service
- to examine the prospects for optimised quality coatings production via the development of expert systems for process control
- to develop an expert system for optimised coatings' design and selection

THE QUALITY ASSURANCE ISSUE IN ADVANCED SURFACE ENGINEERING PRACTICES

In recent years, there has been a continuing and rapid development of advanced surface engineering practices for the optimised enhancement of properties such as corrosion and wear resistance of engineering artifacts. It is now possible to produce coatings and surface treatments of novel composition and microstructure in multilayer format, as appropriate, by a variety of advanced physical and chemical processes, including combination technologies. However, in order for such coatings/treatments to become accepted for us in, for example, manufacturing industry, there is a need to establish, evaluate, and implement appropriate quality assurance procedures.

Although, perhaps, the best known current type of tribological coating is titanium nitride (TiN), there are, in fact, a number of other binary, as well as ternary and quaternary, compounds for potential use as hard coatings. However, TiN technology itself is by no means optimised, especially in the area of quality assurance and reproducibility of properties, indicating a lack of understanding and definition in matters such as stoichiometry, internal stress levels, porosity, adhesion, etc., and the influence of coating conditions and parameters, e.g., bias voltage, argon and nitrogen partial pressures, the nature, condition and geometry of the substrate, etc. [17–20]. Similar problems are evident in the ion plating of other films, e.g., Nb [21,22].

In the context of fast machining or forming operations, it is also of interest to examine the hot hardness for titanium, zirconium, and hafnium mononitrides [23–25]. Machining data [26] indicates the overall superior performance of titanium nitride compared with zirconium nitride and hafnium nitride and, especially, with an uncoated tool, at least with respect to flank wear. On the other hand, crater wear is rather less with zirconium and hafnium nitride coated tools, and the very good performance of zirconium and hafnium carbide coatings has been noted although flank wear was relatively very high.

This variability in performance, and especially its uncertain relationship to room temperature hardness, traditionally taken as a guide to wear resistance, is worrying in the context of the prediction of tool materials with optimum, reproducible properties. Much research is clearly needed in this area.

It is also interesting to consider other coating materials—ternary and quaternary compounds—derived from titanium nitride. A number of ternary compounds [27–32] is currently of considerable interest—compounds where the elements vanadium, aluminium, hafnium, or niobium have been partially substituted for titanium.

It is pertinent to comment on the potential offered by materials of even more complex composition—quaternary compounds such as TiAlVN [28]. Here it is possible to vary the aluminium or vanadium content, each producing materials with grossly differing properties. The full potential of all these novel materials has yet to be clearly explored on a systematic basis.

There is considerable ongoing interest worldwide in the application of these types of compound coatings for the enhancement of cutting tool/press tool life/performance/reliability, as part of a composite or multilayer coatings system. Strafford et al. [24] have discussed their potential role in the enhancement of manufacturing efficiency.

For the most effective and efficient use of coated tools, reliability and guaranteed performance are mandatory, especially within the computer-controlled machine tool scenario. At the present time, even the behaviour of tools coated with TiN via PVD technology, representing the "simplest" system, is uncertain. This variability in performance must, ultimately, be related to a number of interacting factors—coating characteristics and their influence on coated artifact performance. Each coating has characteristics or properties (such as composition, structure, morphology, surface roughness hardness, internal stress level, and adherence) influenced by the PVD processing details, all of which may be measured to provide, in theory, a statement as to the full "nature" of such a coating. In reality, the interaction/compatibility of a given coating with a particular substrate (itself possessing a similar portfolio of significant properties) will also be crucial; that is, there is a need to consider the total coating system. Clearly, in order to be able to control quality in a coating or coating system, all these parameters need to be defined.

The question arises, however, as to the real significance or importance of such properties as, for example, hardness. Thus, generally, it is acknowledged that high hardness may be confidently associated with good wear behaviour. However, a better understanding of the significance of hardness, and especially the known variability of this property on deposition conditions, is lacking. If the true significance of coating hardness on machineability, for example, could be quantified, then this property could be confidently quoted as a quality assurance parameter. In reality, hardness will be related to other aforementioned properties such as internal stress, composition, structure, and morphology, again the precise significances of which—real or apparent—need clarification.

Generally, quality assurance must be seen as a central element in any surface engineering scenario—the defined production of a given coating or surface treatment, requiring the choice of a suitable process technology or combination of technologies, with particular measurable characteristics, which, in turn, faithfully impart the desired performance.

The precise nature and significance of the link between characteristics (e.g., composition, structure, morphology, surface roughness, hardness, internal stress level, and adherence and performance (e.g., corrosion and wear resistance) is not well characterised, and a main object of a research programme would be to systematically explore this linkage, considering also experiences in a range of traditional and experimental coatings and surface treatments with the aim of establishing the relative importance of these characteristics – quality assurance parameters – on performance for optimisation. If this importance can be established, then the linkage could be used through an expert system approach to control, in the most effective manner, the coating/surface treatment process parameters in order to obtain reproducible optimum performance demanded by the manufacturer, i.e. the user of the coated/surface treated component.

REFERENCES

1 Doyle, D. *The Mangement of the Business of Surface Engineering, Proc. Int. Conf. on Surface Engineering: Practice and Prospects*, 12–14 March, 1991, University of South Australia, Adelaide, Australia.

2 Strafford, K. N., Datta, P. K. and Gray, J. S., eds. *Surface Engineering Practice – Processes, Fundamentals and Applications*, Ellis Horwood, Chichester, England, 1990.

3 Shreir, L. L., ed. *Corrosion 2: Corrosion Control*, p. 13.3, Newnes, London, 1976.

4 Tomlinson, W. J. and Mayor, J. P. *Surface Engineering*, 4(3):325, 1988.

5 Datta, P. K., Strafford, K. N., Storey, A. and O'Donnell, A. *Materials and Design*, 3:608, 1982.

6 Datta, P. K. Strafford, K. N., Storey, A. and O'Donnell, A. "Influence of Electroless Nickel Coatings on the Fatigue and Wear Behaviour of Mild Steels," in *Coatings and Science Treatment for Corrosion and Wear Resistance*, Chap. 3, p. 46, K. N. Strafford, P. K. Datta and C. G. Googan, eds., Ellis Horwood, Chichester, 1984.

7 Pearlstein, F. "Electroless Plating," in *Modern Electroplating, 3rd Ed.*, F. A. Lowenheim, ed., John Wiley, New York, 1974.

8 Datta, P. K., Strafford, K. N. and Allaway, S. "Corrosion and Mechanical Properties of Electroless Nickel-High Boron Coatings," Chap. 3.4.3, in *Surface Engineering Practice – Processes, Fundamentals and Applications*, K. N. Strafford, P. K. Datta and J. S. Gray, eds., Ellis Horwood, Chichester, 1990.

9 Tulsi, S. S. and Ebdon, P. *Ind. Corr.*, 1(9):16, 1983.

10 Ebdon, P. *Proc. IV Electroless Nickel Conference*, Chicago, 1985.

11 Moorhouse, P. *Tribology International*, 18:139, 1985.

12 *Proc. Institute of Metal Finishing Conference on Zinc Alloy Electrodeposits*, Aston University, October 1988.

13 Strafford, K. N., Datta, P. K. and Gray, J. S. "Surface Engineering – A Conspectus," in *Surface Engineering Practice – Processes, Fundamentals and Applications*,

K. N. Strafford, P. K. Datta and J. S. Gray, eds., Ellis Horwood, Chichester, England, p. 21, 1990.

14 Vaughan, P. *Metallurgia*, 11:435, 1991.

15 Wilks, T. P. and Strafford, K. N. *Awareness and Take-up of Advanced Materials*, report prepared for the Department of Industry, Trade, and Commerce, Canberra, July 1993.

16 UK Department of Trade and Industry. *Profit through Materials Technology.* National Engineering Laboratory, Glasgow, 1987.

17 Strafford, K. N., Datta, P. K. and Gray J. S. "Applications and Performance of Coatings: Some Perspectives and Prospects," Chapter 3.1 in *Surface Engineering Practice—Processes, Fundamentals and Applications*, K. N. Strafford, P. K. Datta and J. S. Gray, eds., Ellis Horwood, London, 1990.

18 Ahmed, N. A. G. *Ion Plating Technology—Developments and Applications.* John Wiley, NY, 1987.

19 Mattox, D. M. *J. Appl. Physics*, 34:2493, 1963.

20 Teer, D. G. *Tribology*, 8:247, 1975.

21 Thornton, J. A. *Ann. Rev. Mater. Sci.*, 7:239, 1977.

22 Datta, P. K., Strafford, K. N., Lin, D. S., Ward, L. P., Hill, R. and Russell, G. J. "Structure and Preferred Orientation in Ion-Plated Nb Films and Correlation of the Substrate Bias Voltage with Calculated Strain Energies," *Thin Solid Films*, p. 168, 1988.

23 Datta, P. K., Ward, L. P., Hill, R. and Strafford, K. N. "The Effect of Deposition Parameters on the Microstructure of Nb Coatings," Chap. 1.2.3, in *Advances in Surface Engineering*, K. N. Strafford, P. K. Datta and J. Gray, eds., Ellis Horwood, Chichester, England, 1990.

24 Strafford, K. N., Datta, P. K. and Gray, J. "Protective Coatings: Some Perspectives and Prospects," in *Surface Engineering Practice—Processes, Fundamentals and Applications*, Chap. 3.1, K. N. Strafford, P. K. Datta and J. Gray, eds., Ellis Horwood, Chichester, England, 1990.

25 Strafford, K. N. *Corrosion Science*, 19:49, 1979.

26 Quinto, D. T., Wolfe, G. J., and Jindal, P. C. *Thin Solid Films*, 153:19, 1987.

27 Sproul, W. D. *Proc. 14th Int. Conf. on Metallurgical Coatings*, San Diego, March 1987, *Surface and Coatings Technology*, 33:133, 1987.

28 Boelens, S. and Veltrop, H. *Proc. 14th Int. Conf. on Metallurgical Coatings*, San Diego, March 1987, *Surface and Coatings Technology*, 33:63, 1987.

29 Konig, U. *Proc. 14th Int. Conf. on Metallurgical Coatings*, San Diego, March 1987, *Surface and Coatings Technology, Thin Solid Films*, 33:91, 1987.

30 Hakansson, G. and Sundgren, J. E. *Proc. 14th Int. Conf. on Metallurgical Coatings*, San Diego, March 153:159, 1987.

31 Fenske, G. R. and Kaufherr, N. *Proc. 14th Int. Conf. on Metallurgical Coatings, Thin Solid Films*, 153:159, 1987.

32 Knotek, O., Leyendecker, T. and Jungblut, F. *Proc. 14th Int. Conf. on Metallurgical Coatings, Thin Solid Films*, 153:83, 1987.

PART I

COATINGS AND SURFACE MODIFICATIONS—PROCESS TECHNOLOGIES

CHAPTER 1

Plasma Sprayed Ceramic Coatings

R. McPHERSON[1]*

1.1 INTRODUCTION

"ADVANCED" ceramics have found increasing industrial use for critical applications in recent years because of their unique thermal, mechanical, chemical, electrical, magnetic, and optical properties. Ceramics are brittle, refractory materials and must generally be manufactured by sintering of powder compacts at high temperatures. Machining operations are difficult and normally require diamond tools. Size and shape limitations imposed by conventional processing methods and the high cost, therefore, restrict many potential applications. An alternative approach is to use ceramic coatings on other, usually metallic, materials to provide the required special properties at lower cost and in configurations unattainable by other means. Plasma spraying is the most widely used method for preparing ceramic coatings, but coating properties are quite different to conventional ceramics and not well understood. This chapter reviews the properties and applications of plasma-sprayed ceramic coatings and examines the relationship between their properties, microstructure, and process parameters.

1.2 THE PLASMA-SPRAYING PROCESS

Metal spraying has been used commercially since the 1930s, employing

*Deceased 9th November, 1993.

[1]CSIRO Division of Manufacturing Technology, Locked Bag No. 9, Preston, Victoria 3072, Australia.

processes in which wires or powders are melted in a combustion flame and the resultant stream of molten particles is directed onto a cool substrate where a coating is formed by the particles' successive impact and solidification. Ceramic coatings were introduced in the 1950s using similar techniques. For example, in the Norton "Rokide" process [1] a sintered ceramic rod was introduced axially into an oxygen-acetylene flame, whereas other companies developed equipment employing injection of ceramic powders into an oxy-acetylene or oxy-hydrogen flame [2]. A limitation of these processes is that the flame temperature is relatively low compared with the high melting points of many of the ceramics of interest such as aluminium oxide $(Al_2O_3)-2050°C$ and zirconium oxide $(ZrO_2)-2750°C$, and it is therefore difficult to produce a high velocity stream of completely molten particles. Development in the 1960s of the direct current plasma torch, with a jet temperature of around $10,000°C$ and a velocity of several hundred metres per second, provided a means of producing a high-velocity stream of molten particles of any material that did not vapourise excessively or dissociate [3]. This has led to the widespread industrial application of ceramic coatings and continuing development of new materials and applications.

The plasma jet is produced by a direct current arc between a tungsten cathode and an annular water cooled copper anode in a chamber into which an inert gas (argon, helium, nitrogen, hydrogen, or a mixture of these) is injected. Powder, carried in a stream of carrier gas, is usually injected radially into the jet where it is entrained, accelerated, and melted. Powder characteristics are critical, for if the particle size is too small, vapourisation may be a problem and if it is too large, complete melting may not be possible in the short residence time (a few milliseconds) within the plasma. Good powder flowability, achieved by a narrow particle size distribution and equiaxed particle shape, is also important to ensure controlled feeding into the plasma. It is also an advantage to use dense powder particles to achieve rapid melting. Ideally, all of the injected particles are completely melted, and on impact with a substrate, they spread and rapidly solidify to form a dense deposit with a lamellar microstructure [4].

Plasma spraying is usually conducted in air (APS), but this results in the partial oxidation of metallic materials and incorporation of oxide particles within the coating. Spraying in an inert atmosphere overcomes the problem, especially for reactive metals, and is commonly carried out at reduced pressure, referred to as low pressure plasma spraying (LPPS) or vacuum plasma spraying (VPS). Although LPPS is not necessary for oxide ceramics from the oxidation point of view, some studies have been conducted on LPPS ceramics, which show evidence that superior properties may be obtained under certain conditions [5].

1.3 THE STRUCTURE OF PLASMA-SPRAYED COATINGS [6]

When a spherical liquid droplet strikes a flat surface at high velocity, it flattens to a disc, but the radially flowing thin sheet of liquid becomes unstable and disintegrates at the edges into small droplets. In the plasma-spraying case, the substrate is at a temperature much lower than the melting point of the droplet, and the heat transfer rate to the substrate is very high so that spreading and breakup of the droplet is interrupted by solidification. The mean initial droplet diameter is ~50 μm for practical reasons, and they usually form solidified discs with a ~150-μm diameter and are ~3-4-μm thick. The estimated cooling rate of these discs is 10^6 Ksec^{-1}, that is, equivalent to "splat quenched" material. Classical heat transfer considerations suggest that the interface between adjacent particles consists of small distributed contact points within an effectively nonconducting film arising from entrapped gas. The microstructure of the "ideal" plasma-sprayed coating is thus related to the coating formation process through two predominant factors, the nature of the interface between randomly stacked disc shaped lamellae and the internal structure within these lamellae produced by their rapid solidification (Figure 1.1).

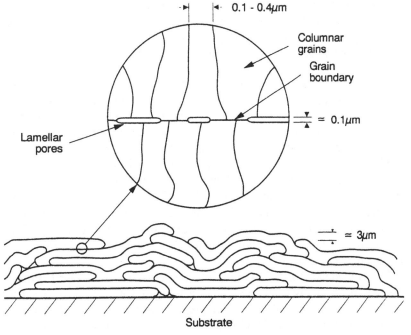

Figure 1.1 Diagrammatic representation of coating structure.

Deviations from the ideal behaviour occur in real coatings because of the incorporation of partly melted particles, splashing of droplets on impact, and defects arising from faults in the spraying process such as fluctuations in powder feed rate.

1.3.1 COATING POROSITY

Plasma-sprayed deposits have porosities ranging from a few percent up to about 20%, depending upon powder characteristics and processing conditions. Studies of the pore size distribution of alumina coatings, by the mercury intrusion method (MIP), suggest that the predominant pore size is ~0.1 mm [7]. The MIP technique, however, gives misleading results if the pore substructure consists of large pores interconnected by narrow channels since the pore diameter measured is that of the channels and the fraction porosity is determined by the total connected pore volume. Other techniques must therefore be used to obtain a complete description of pore size distribution. The pore size distribution of Ni-Al plasma-sprayed coatings has been studied using a combination of MIP and quantitative analysis of scanning electron microscope (SEM) images [8]. This approach provided independent estimates of the pore size distribution at larger pore diameters (>1 μm). The results were consistent with a two-level porosity model consisting of isolated 1-μm to 10-μm pores interconnected by a network of ~0.1-μm planar pores. Differences in pore microstructure between various coating types could be used to explain differences in coating strength.

The extremely fine porosity observed in plasma-sprayed coatings has been interpreted, in part, as incomplete contact between lamellae, as referred to previously. This view is supported by transmission electron microscopy of transverse sections of alumina coatings, which showed narrow planar pores (~0.1 μm) between lamellae [9].

SEM examination of the surfaces of individual lamellae of thermally sprayed ceramic coatings also shows a network of fine cracks (~0.1-μm wide) perpendicular to the coating plane. These are formed by stresses arising from restraint of thermal contraction of the lamellae by underlying material during cooling from the solidification temperature [6]. Such cracks are not formed in metallic coatings because the thermal strain is accommodated by plastic flow.

Direct evidence for the microporous model of plasma-sprayed ceramic coatings has recently been provided by a study in which the pore structure, decorated by a copper plating technique, was quantitatively analysed from atomic number contrast SEM images [10]. This revealed the orthogonal pattern of planar interlamellar pores and intralamellar microcracks

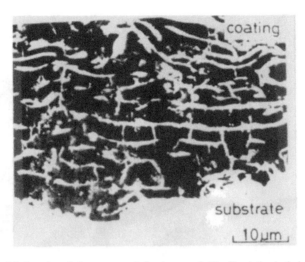

Figure 1.2 Porosity of plasma-sprayed alumina revealed by Cu plating technique [10].

(Figure 1.2). The real area of contact between lamellae ("bonding rate") was found to be 20–30%, close to that postulated earlier from elastic modulus data. This model therefore postulates that "ideal" plasma-sprayed ceramic coatings contain an intrinsic intersecting network of planar pores, approximately 0.1-μm wide. Additional, larger pores occur in "real" coatings because of defects in the coating process.

1.3.2 METASTABLE PHASES

Extended solid solubility beyond the equilibrium limits is commonly observed in the case of alloy systems. Plasma-sprayed Al_2O_3–5 wt% Y_2O_3 consists of a metastable spinel solid solution, but on heat treatment, the equilibrium structure of α-Al_2O_3 plus $Y_3Al_5O_{12}$ is formed [16]. There is, however, a theoretical limit beyond which an extended solid solution cannot form in rapidly cooled alloys, for thermodynamic reasons, and extremely fine grain size duplex structures are formed. For example (Figure 1.3), Al_2O_3–25 wt% ZrO_2 coatings consist of an ultrafine dispersion of γ-Al_2O_3 and tetragonal ZrO_2 [17].

Another interesting example of a metastable binary coating is plasma-sprayed zircon ($ZrSiO_4$), which consists of a mixture of tetragonal and monoclinic ZrO_2, the proportions depending on spraying conditions, in silica glass (Figure 1.4). The complex microstructure observed is influenced by a metastable miscibility gap in the system [18].

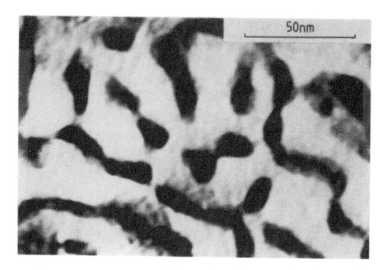

Figure 1.3 Transmission electron micrograph of Al_2O_3–25 wt% ZrO_2 plasma-sprayed coating [17].

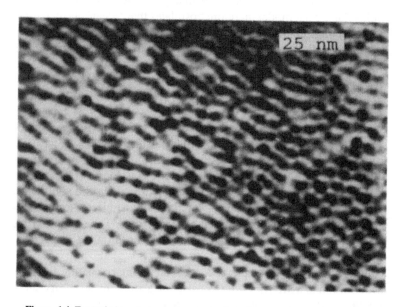

Figure 1.4 Transmission electron micrograph of plasma-sprayed zircon coating [18].

8

1.3.3 THE STRUCTURE OF LAMELLAE

Each of the lamellae, of which the coating is made up, solidifies as a separate entity so that the internal structure of each may be considered separately. As pointed out previously, each particle cools at a rate comparable to rapid solidification processes, such as splat quenching and melt spinning, and they therefore have different structures than ceramics prepared by sintering. Crystallisation of a liquid occurs by nucleation and growth, but if the cooling rate is such that no nuclei are formed in the time interval between the melting point and the glass transition temperature, the liquid will harden to form an amorphous solid or glass. This is observed, for example, in some Al_2O_3-ZrO_2 alloys [11].

Metastable crystalline phases can also form directly from the liquid if it is rapidly cooled. The best known example of this is plasma-sprayed alumina, which consists predominately of metastable γ-Al_2O_3 rather than α-Al_2O_3, the only stable crystalline phase [12]. The reason for this effect is that the phase that nucleates from an undercooled melt is not necessarily the one with the lowest free energy (the equilibrium phase), but one with the lowest energy barrier to nucleation. If the phase that nucleates happens to be metastable, it may be retained to ambient temperature if the cooling rate is sufficiently rapid to prevent transformation to the stable phase [13]. Alumina coatings often contain some α-Al_2O_3, but this can be explained by the incorporation of incompletely fused powder, and the fraction α-Al_2O_3 is thus a function of the spraying conditions [14].

The crystal size and morphology within lamellae depend upon their relative nucleation and growth rates as a function of temperature. In general, high cooling rates give rise to high nucleation rates, and the lamellae are therefore polycrystalline with crystal sizes much smaller than conventionally processed materials. In most cases, it seems that rapid nucleation occurs at the cooler surface of the flattened droplet at large undercooling, and crystals grow rapidly to form a columnar grain structure (Figure 1.5). Rapid crystal growth leads to rapid evolution of the heat of fusion, an increase in temperature and suppression of further nucleation. On the other hand, if the crystal growth rate is relatively low, further nucleation occurs in the undercooled liquid, and a very fine equiaxed structure is formed.

1.4 PROPERTY-MICROSTRUCTURE RELATIONSHIPS

The completely different structure of plasma-sprayed ceramic coatings, compared with their conventionally processed counterparts, naturally leads to major differences in physical properties. The microporous sub-

Figure 1.5 Fracture surface of plasma-sprayed zirconia-ceria coating showing columnar grain structure within lamellae [15].

structure, in particular, dominates many of their properties as discussed below.

1.4.1 ELASTIC MODULUS

The Young's modulus (E) of dense α-Al$_2$O$_3$ is ~400 GPa and in sintered form with 10% porosity ~300 GPa, whereas the reported modulus of plasma-sprayed alumina is only 30–90 GPa. Recent measurements parallel and perpendicular to the coating plane have shown pronounced elastic anisotropy with E in the coating plane of 89 GPa and E perpendicular to it of 29 GPa [19]. Furthermore, nonlinear elastic behaviour in tension but linear elastic behaviour in compression was observed perpendicular to the coating plane, whereas the behaviour within the plane was linear elastic in both tension and compression (Figure 1.6).

Nonlinear stress-strain behaviour has also been observed in alumina coatings under bending [20]. Part of the modulus reduction, compared with sintered alumina, may be ascribed to the fact that coatings consist predominately of γ-Al$_2$O$_3$, but the major part, and the anisotropy, must be related to the micro-porosity distribution. The low modulus perpendicular to the coating plane can be simply related to the fraction real interlamellar contact area of 20–30%, and the nonlinear elastic behaviour to the opening of interlamellar pores. Higher elastic modulus in the coating plane may be related to the jigsaw puzzle-like structure arising from the thermally induced microcracks in this plane.

1.4.2 FRACTURE TOUGHNESS

It has been stated that the Young's modulus of coatings is much lower than that of the bulk material because of the interlamellar porosity, and it would also be expected that the effective fracture surface energy (γ_{eff}) would be lower in the coating plane since a crack would pass from one region of good contact to another, and the effective fracture surface area would therefore be lower. Thus γ_{eff} would be anticipated to be approximately equal to $a_r\gamma_o$, where a_r is the fraction real area of contact between lamellae and γ_o the fracture surface energy of the dense solid. It would be surprising, however, if fracture occurred by propagation of a single crack in such a heterogeneous structure. Crack branching would be more likely leading to an increase in the crack surface area by a multiplying factor (m) so that: $\gamma_{eff} = ma_r\gamma_o$.

Measurements of the critical strain energy release rate ($G_c = 2\gamma_{eff}$) of

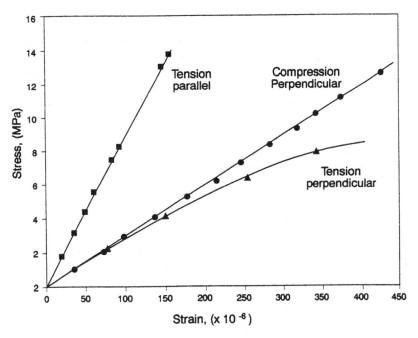

Figure 1.6 Stress-strain curves for plasma-sprayed alumina parallel and perpendicular to the coating plane [19].

plasma-sprayed coatings, using the double cantilever beam technique [21–24], showed a wide scatter of values, together with variations of crack length, which may be related to changes in the degree of crack branching. The critical stress intensity factor (K_{IC}), which is related to both E_{eff} (effective Young's modulus) and γ_{eff}, is also strongly influenced by microstructure. Typical values for zirconia-based coatings are around 1 MPa m$^{1/2}$ [24,25]. Attempts to increase the toughness of zirconia coatings by a transformation toughening effect resulted in a mean K_{IC} of about 1.4 MPa m$^{1/2}$ [24]. This should be compared with typical K_{IC} values (MPa m$^{1/2}$) for conventional ceramics of approximately 1 for porcelain, 3–4 for alumina, and 4–12 for transformation toughened zirconia [26].

1.4.3 STRENGTH AND ADHESION

The "strength" of coatings is usually determined by a test such as ASTM C 633-79, in which the coating on one end of a 25-mm diameter steel stub is attached to a similar stub by an epoxy adhesive, and the composite specimen is loaded in tension to failure normal to the coating surface. Fracture may occur in four ways: through the coating itself (cohesive failure), at the coating substrate interface (adhesive failure), through the adhesive layer, or by a combination of these paths. The test is mainly used for quality control and provides a relatively simple comparative means of determining the tendency for the coating to delaminate or come off the substrate in service. The test is subject to considerable variability and is influenced by specimen preparation details and loading conditions, as well as inherent coating properties. Under ideal testing conditions, the strength is influenced by microstructure through the materials parameter, fracture toughness (K_{IC}), and the dimensions of the largest crack-like defect present (c): $\sigma_f = K_{IC}(Yc)^{-1/2}$, where Y is a geometric factor. As discussed above, K_{IC} is controlled by the pore substructure, which therefore tends to be the dominant microstructural factor influencing the strength of ceramic coatings [27].

The mechanism of coating adhesion is controversial, however. Whatever the nature of the bonding forces between the individual coating lamellae and the substrate, the interface probably consists of regions of good contact between regions of ineffective contact as proposed for the interlamellar structure [6]. In practical terms, substrate preparation is important, and a freshly grit-blasted surface is essential to achieve good adhesion. It has also been observed that the adhesion of ceramic coatings to metal substrates is improved considerably if a metallic "bond coat," such as Ni-Al, is used between the ceramic and substrate. Again, the reason for this effect is obscure, although measurements of the fracture surface energy for cohesive failure of alumina coatings and adhesive failure between alumina and a steel substrate and Ni-Al subcoat on steel showed an increase in energy in the order: adhesive on steel, cohesive, adhesive on bond coat. It was suggested that the increased fracture energy for the alumina-bond coat interface compared with the alumina-substrate interface, arises from an additional contribution due to plastic deformation of subcoat lamellae, an effect which is not possible for failure at the ceramic-substrate interface. The higher "toughness" for ceramic-bond-coat fracture is manifested as improved adhesion [6,21].

1.4.4 THERMAL CONDUCTIVITY

An important application of plasma-sprayed ceramics is as thermal barrier coatings, which depend upon the establishment of a steep temperature gradient through a refractory oxide coating on a metallic substrate in an elevated temperature environment. The operating temperature of the metallic component is reduced by this means. Thermal conductivity is clearly a significant factor in the choice of coating material, and zirconium oxide–based coatings are commonly used because of their low intrinsic thermal conductivity, "refractoriness," and relatively high thermal expansion coefficient, which matches that of many metallic substrates [28]. Measurements of the thermal conductivity of zirconia plasma-sprayed coatings have shown, in fact, that the thermal conductivity and thermal diffusivity are about one-quarter those of similar materials in sintered form [29]. The reason for this can, again, be traced to the lamellar microstructure of coatings and the imperfect contact between lamellae [30]. Thus, conduction of heat in a vacuum is limited normal to the coating plane to a path of series-parallel contact regions between lamellae. In a gaseous environment, conduction is also possible through the parallel paths of the interlamellar pores, but the very small width of these (~ 0.1 μm) is comparable to the mean free path of the gas molecules and the conduction through them is severely limited. This model gives very good agreement with measured values as shown in Table 1.1.

TABLE 1.1. Measured [29] and Calculated [30] Values of Thermal Conductivity of Plasma-Sprayed ZrO_2-CaO Coatings in Various Atmospheres.

Atmosphere	Thermal Conductivity (W m^{-1}K^{-1})	
	Experimental	Calculated
Vacuum	0.45	0.55
Argon	0.68	0.65
Helium	0.82	0.80

1.4.5 ELECTRICAL AND MAGNETIC PROPERTIES

Specific electrical-magnetic properties of sintered ceramics have led to their widespread commercial application, and plasma spraying has been employed to prepare similar materials in the form of thick films. Again, microstructural features arising from the coating process (metastable phase formation, fine grain size, porosity, and pore geometry) have major effects on properties. The dielectric constant (x) is a function of the intrinsic constant of the material, as well as the fraction porosity and the pore geometry. Alumina plasma-sprayed coatings may be used as a dielectric, but it is observed that the value of x is up to thirty-five, even after drying at 135°C, whereas the intrinsic value for sintered α-Al_2O_3 is only ten [31]. The reason for this appears to be the absorption of water ($x \sim 100$) by the porous, reactive γ-Al_2O_3 phase, an effect which may be overcome by sealing the coating using an epoxy filler immediately after spraying [32].

Dielectric loss (tan δ) is an intrinsic property of the material and may be strongly influenced by the presence of certain impurities. For this reason high purity alumina is commonly used for sprayed dielectric layers. The dielectric strength of alumina coatings is reduced to 100–200 kV cm^{-1} compared with 700 kV cm^{-1} for sintered alumina because of its inherent porosity [32]. This reduction is greater than that expected for a model based on isolated pores, another important effect of the continuous network of interlamellar pores and intralamellar microcracks within thermally sprayed ceramic coatings on properties.

Electrical resistors can also be prepared in quantity at low cost by plasma spraying. Thus Fe_3O_4-NiO and Fe_3O_4-Al_2O_3 mixed powders have been sprayed to prepare deposits with controlled electrical resistance, which consists of a dispersion of conducting particles in an insulating matrix [33]. Similar cermet coatings (Al_2O_3 or ZrO_2 plus W, Cr, Ni, or Mo) may be used as electrical heaters at power ratings of 80 kW m^{-2} at temperatures below 300°C [34]. An advantage of this technique is the ability to produce heaters of particular shapes by masking. An example of the application of this approach to a commercial product is an electrically heated roller for photocopiers, in which a stainless steel roller was sprayed successively with a Ni-Al-Mo bond coating, spinel ($MgAl_2O_4$) insulating layer, and cermet resistance heat layer consisting of Cu-Zn ferrite, TiO_2-Ni,Cr or Al_2O_3-Ni,Cr [35].

A rather different use of plasma-sprayed ceramic conductors is the porous $LaCrO_3$ interconnection, Ni-YSZ fuel electrode, and $La_{1-x}Sr_xMnO_3$ air electrode in solid oxide fuel cells based on a ZrO_2 solid electrolyte [36]. Attempts to produce a solid oxide fuel cell in which all of the elements are thermally sprayed have not yet been successful because of the problem of gas permeability of the yttria stabilised zirconia electrolyte

[37]. Plasma spraying may, however, be used to prepare disposable oxygen sensors that are directly immersed into molten steel to measure the oxygen content [38]. The pin sensor consists of a Mo wire coated with a Cr-Cr_2O_3 reference layer, followed by a fully stabilised ZrO_2-Y_2O_3 solid electrolyte layer 0.4-mm thick and a partial Al_2O_3 coating. These low-cost sensors have excellent thermal shock resistance and faster response time than conventional sensors.

The preparation of magnetically soft ceramic ferrites by plasma spraying has been reported [39]. Poor magnetic properties were obtained in the as-sprayed condition, but heat treatment produced significant improvements, probably due to sintering and crystal growth to produce a structure similar to that of a conventional ceramic.

The recent discovery of ceramic superconductors with critical temperature above the boiling point of nitrogen has resulted in an enormous interest in their processing, including the use of thermal spraying. As-sprayed deposits of $YBa_2Cu_3O_x$ have a metastable, oxygen deficient, simple cubic structure, which is not superconducting, but they can be transformed to the superconducting orthorhombic phase with sharp transition at ~ 90 K by heat treatment at $\sim 900°C$ in air or oxygen followed by slow cooling [40]. A high critical current density, J_c, is required for most applications, and the relatively low value of ~ 200 amp cm^{-2} achieved [41] is disappointingly low.

1.4.6 CORROSION AND OXIDATION

Ceramics are generally very corrosion-resistant and oxidation-resistant at high temperatures, but problems arise with their application in the form of plasma-sprayed coatings on metallic substrates because of connected porosity. The use of plasma-sprayed ceramics as wear-resistant seals in pumps and similar applications therefore often requires the use of a corrosion-resistant substrate [41], but materials that may be satisfactorily uncoated may give problems when coated with alumina or chromia because of crevice corrosion and a less susceptible substrate may be necessary [42].

Similarly, the metallic substrate may oxidise at high temperatures under ceramic thermal barrier coatings because of the diffusion of oxygen through the interconnected pores, and the substrate must therefore have adequate oxidation resistance. This effect is a significant factor in the service failure of thermal barrier coatings in gas turbine applications [43]. Although oxide ceramics will withstand air and normal combustion products at high temperatures, they may suffer corrosion by impurities in the gas phase. Zirconia-yttria thermal barrier coatings, for example, may preferentially lose Y_2O_3 by reaction with vanadates, salt, and other contaminants, resulting in destabilisation of the structure and failure under thermal cycling conditions. The use of other stabilising oxides, such as scandium oxide, reduces this effect, but at a price [44].

1.5 INDUSTRIAL USE OF CERAMIC PLASMA-SPRAYED COATINGS

The bulk of plasma-sprayed ceramic coatings presently used industrially exploit the wear resistance of alumina or alumina-titania alloys (5–40 wt% TiO_2) and chromia. They are employed in a wide variety of applications in the chemical engineering, textile, and paper industries—in particular, as pump liners, impellers and sleeves, guides, rollers, and drums. Although the hardness and strength of these coatings are not as high as the equivalent sintered material because of their microstructure, as discussed previously, their unique structure provides considerable strain tolerance and significant resistance to mechanical damage.

A significant special purpose application is large inking rollers in the printing industry, which are plasma-sprayed with chromium oxide and laser-embossed. Plasma-sprayed ZrO_2–25% Y_2O_3 on cooling rolls in steel sheet continuous annealing lines have proved successful in extending life by a factor of eight compared with electroplated chromium because of superior wear, thermal shock, and surface heat transfer properties [46].

Thermal barrier coatings, predominately ZrO_2–6 to 8 wt% Y_2O_3 (tetragonal) and so-called magnesium zirconate, ZrO_2–20 wt% MgO (cubic), are extensively used in gas turbines to allow the use of higher temperature combustion and, hence, better engine performance. They are used on the burner cans and guide vanes over a Ni-Co-Cr-Al-Y LPPS subcoat. These coatings would also be desirable on the rotating blades, but their reliability is apparently not satisfactory. It now seems that electron beam physical vapour-deposited coatings offer better performance, and their high cost may be acceptable in this application [47]. Zirconia-based thermal barrier coatings have also been used in the combustion chambers of diesel engines to improve thermal efficiency, but again, reliability is a problem. A lower cost system employing a ZrO_2-CeO_2-Y_2O_3 coating on APS Ni-Cr-Al-Y subcoat shows promise of adequate performance for the piston crown, cylinder head, valve face, and throat and exhaust ports [48].

Several industrial applications exploit the electrical properties of ceramic coatings, mostly involving the use of alumina as a low-cost dielectric or insulating thick film. A typical high-volume use in an automotive electrical component employs sprayed alumina insulating pads under sprayed copper pads onto which electrical components are soldered. This is an example of an important advantage of plasma spraying, the ability to coat specific regions by masking and the production of laminated structures.

LPPS offers superior coating properties in some cases, but the high cost would seem to be a major disadvantage in many industrial situations. Improvements in commercial LPPS-spraying systems, however, will probably lead to more widespread use of the process for special ceramic coating applications [49].

Although plasma-sprayed ceramic coatings are widely used in many industries, little advantage has been taken so far of a unique feature of the process, namely, the formation of metastable phases and extended solid solutions, which could result in the preparation of materials with intrinsic properties quite different to conventionally processed ceramics. This, combined with the ability to prepare complex composites by plasma spraying, offers opportunities for development of completely new products.

1.6 CONCLUSIONS

Plasma-sprayed ceramic coatings have unusual microstructures resulting in properties that differ considerably from their sintered counterparts. The elastic modulus, fracture toughness, fracture strength, thermal conductivity, and dielectric strength, in particular, are inferior to the sintered material. On the other hand, the microstructure results in considerable strain tolerance and resistance to thermal shock.

The unique properties of plasma-sprayed ceramic coatings have led to their widespread commercial use, particularly for wear-resistant surfaces, thermal barriers, and dielectrics. Most of the present applications make use of a limited range of materials based on alumina, chromia, and zirconia.

There have been many reports of specialised applications of plasma-sprayed ceramics, and more widespread industrial use can be expected as the technology develops and property-structure relationships are better understood.

1.7 REFERENCES

1 Riley, M. W. "New Flame Sprayed Ceramics," *Mater. Methods,* 42:96–98, 1955.

2 Bliton, J. C. and Rechter, H. L., "Determination of Physical Properties of Flame-Sprayed Ceramic Coatings," *Bull. Amer. Ceram. Soc.,* 40:683–688, 1961.

3 Mash, D. R., Weare, N. E. and Walker, D. L. "Process Variables in Plasma Jet Spraying," *J. Metals,* 13:473–478, 1961.

4 Fauchais, P., Bourdin, E., Coudert, J-F. and McPherson, R. "High Pressure Plasmas and Their Application to Ceramic Technology," *Topics in Current Chemistry,* Springer Verlag, Berlin, 107:59–183, 1983.

5 Sasazaki, K., Miyamoto, Y. and Koizumi, M. "Effect of LPPS on Mechanical

Properties of Al_2O_3 Coatings and Examination of Optimum Spraying Conditions," *J. Ceram. Soc. Japan,* 95:1175–1180, 1987.

6 McPherson, R. "The Relationship between the Mechanism of Formation, Microstructure and Properties of Plasma Sprayed Coatings," *Thin Solid Films,* 83:297–310, 1981.

7 Vardelle, M. and Besson, J. L. "γ-Alumina Obtained by Arc Plasma Spraying: A Study of the Optimisation of Spraying Conditions," *Ceram. Int.,* 7:48–54, 1981.

8 McPherson, R. and Cheang, P. "Microstructural Analysis of Ni-Al Plasma Sprayed Coatings," *Proc. 12th Int. Thermal Spraying Conf.,* London, 1989, I. A. Bucklow, ed., Abington Pub., Cambridge, 1:237–246, 1990.

9 McPherson, R. and Shafer, B. V. "Interlamellar Contact within Plasma-Sprayed Coatings," *Thin Solid Films,* 97:201–204, 1982.

10 Arata, Y., Ohmori, A. and Li, C-J. "Electrochemical Method to Evaluate the Connected Porosity in Ceramic Coatings," *Thin Solid Films,* 156:315–325, 1988.

11 Krauth, A. and Meyer, H. "On Modifications Produced by Chilling and on Their Crystal Growth in Systems Containing Zirconia," *Der Deutsch. Keram. Ges.,* 42:61–72, 1965.

12 Ault, N. N. "Characteristics of Refractory Oxide Coatings Produced by Flame-Spraying," *J. Amer. Ceram. Soc.,* 40:69–74, 1957.

13 McPherson, R. "Formation of Metastable Phases in Flame and Plasma Prepared Alumina," *J. Mater. Sc.,* 8:851–858, 1973.

14 McPherson, R. "On the Formation of Thermally Sprayed Alumina Coatings," *J. Mater. Sc.,* 15:3141–49, 1980.

15 Heintze, G. N. and McPherson, R. "Structure of Plasma-Sprayed Zirconia Coatings," *Advances in Ceramics, 24:431*–438, 1988.

16 McPherson, R. Unpublished work.

17 McPherson, R., "Microstructure of Thermally Sprayed Coatings," *Proc. Intl. Symp. Advanced Thermal Spraying Technology and Allied Coatings,* High Temp. Soc. Japan, Osaka, Japan, pp. 295–300, 1988.

18 McPherson, R. "Spherulites and Phase Separation in Plasma Dissociated Zircon," *J. Mater. Sc.,* 9:2696–2704, 1984.

19 McPherson, R. and Cheang, P. "Elastic Anisotropy of APS Alumina Coatings," *Proc. 7th. CIMTEC World Ceramics Conf.,* Italy, 1990 (in press).

20 Shi, K-S., Qian, Z-Y. and Zhuang, M-S. "Microstructure and Properties of Sprayed Ceramic Coating," *J. Amer. Ceram. Soc.,* 71:924–929, 1988.

21 Berndt, C. C. and McPherson, R. "The Adhesion of Plasma Sprayed Ceramic Coatings to Metals," *Matls. Sc. Research,* J. Pask and A. Evans, eds., New York, 14:619–628, 1981.

22 Ostojic, P. and McPherson, R. "Determining the Strain Energy Release Rate of Plasma Sprayed Coatings Using a Double Cantilever Beam Technique," *J. Amer. Ceram. Soc.,* 71:891–899, 1988.

23 Heintze, G. N. and McPherson, R. "Fracture Toughness of Plasma-Sprayed Zirconia Coatings," *Surface and Coating Tech.,* 34:15–23, 1988.

24 Heintze, G. N. and McPherson, R. "A Further Study of the Fracture Toughness of Plasma Sprayed Zirconia Coatings," *Surface and Coatings Tech.,* 36:125–132, 1988.

25 Kleer, G., Schonholz, R., Doll, W., Sturlese, S. and Zacchetti, N. "Interface

Crack Resistance of Zirconia Base Thermal Barrier Coatings," *Proc. 7th CIMTEC World Ceramics Conf.*, Italy, 1990 (in press).

26 Ashby, M. F. and Jones, D. R. H. *Engineering Materials 2*, Pergamon Press, Oxford, p. 150, 1986.

27 McPherson, R. "A Review of Microstructure and Properties of Plasma Sprayed Ceramic Coatings," *Surface and Coatings Tech.*, 39/40:173–181, 1989.

28 Strangman, T. E. "Thermal Barrier Coatings for Turbine Aerofoils," *Thin Solid Films*, 127:93–105, 1985.

29 Wilkes, L. E. and Lagerdrost, J. F. "Thermophysical Properties of Plasma Sprayed Coatings," NASA Contract Report CR 121144, National Aeronautics and Space Administration, 1973.

30 McPherson, R., "A Model for the Thermal Conductivity of Plasma-Sprayed Ceramic Coatings," *Thin Solid Films*, 112:89–95, 1984.

31 Brown, L., Herman, H. and MacCrone, R. K. "Plasma-Sprayed Insulated Metal Substrates," *Proc. 11th Int. Conf. Thermal Spraying*, Montreal, 1986, Pergamon Press, New York, pp. 507–513.

32 Pawlowski, L. "The Relationship between Structure and Properties in Plasma-Sprayed Alumina Coatings," *Surface and Coating Technol.*, 35:286–298, 1988.

33 Smyth, R. T. and Anderson, J. C. "Production of Resistors by Arc Plasma Spraying," *Electro Component Sc. and Tech.*, 7:135–145, 1980.

34 Taylor, A. J., Parker, N. J. and Wellhofer, F. "Plasma Sprayed Cermets: Their Electrical Resistivity and Thermal Stability," *Proc. 12th Int. Thermal Spraying Conf.*, London, 1989, I. A. Bucklow, ed., Abington Pub., Cambridge, 2:257–266, 1990.

35 Shibata, R., Kaskoshi, T., Iimura, T. and Harada, H. "Surface Heating Fuser Roll Made by Arc Plasma Spraying," *Proc. Intl. Symp. Advanced Thermal Spraying Technology and Allied Coatings*, High Temp. Soc. Japan, Osaka, Japan, pp. 235–239, 1988.

36 Okai, R., Yoshida, S., Kaji, I., Hasegawa, M., Yamanuchi, H. and Nagata, M. "Application of Plasma Spraying Process for Porous Electrodes," *Proc. Intl. Symp. Solid Oxide Fuel Cells*, Nagoya, Japan, pp. 115–122, 1989.

37 Kasuga, Y. Nagata, S. and Hayashi, K. "Thermal Spraying for Solid Oxide Fuel Cell," *Proc. Intl. Symp. Advanced Thermal Spraying Technology and Allied Coatings*, High Temp. Soc. Japan, Osaka, Japan, pp. 247–252, 1988.

38 Urata, K., Ogura, T., Matsuoka, M. Ohmori, A. and Arata, Y. "Application of Ceramic-Spray Technique to Manufacture of Oxygen Sensor for Steelmaking," *Proc. Intl. Symp. Advanced Thermal Spraying Technology and Allied Coatings*, High Temp. Soc. Japan, Osaka, Japan, pp. 241–246, 1988.

39 Preece, I. and Andrews, W. D. "Plasma Spraying of Ferrites," *J. Mater. Sc.*, 8:964–967, 1973.

40 Heintze, G. N., McPherson, R., Tolino, D. and Andrikidis, C. "The Structure of Thermally Sprayed $YBa_2Cu_3O_{7-x}$ Superconducting Coatings," *J. Mater. Sc. Lett.*, 7:251–253, 1988.

41 Pawlowski, L., Gross, A. and McPherson, R. "Microstructure of Plasma Sprayed $YBa_2Cu_3O_x$ Superconducting Coatings," *J. Mater. Sc.* (in press).

42 Fukumoto, M., Wada, Y., Umemoto, M. and Okane, I. "Effect of Connected Pores on the Corrosion Behaviour of Plasma Sprayed Alumina Coatings," *Surface and Coatings Tech.*, 39/40:711–720, 1989.

43 Askay, A. A. and Tucker, R. C. "Electrochemical Corrosion Studies of Alloys Plasma Sprayed with Cr_2O_3," *Surface and Coatings Tech.*, 39/40:701–709, 1989.

44 Miller, R. A. "Current Status of Thermal Barrier Coatings—An Overview," *Surface and Coatings Tech.*, 30:1–11, 1987.

45 Jones, R. L. "Scandia-Stabilised Zirconia for Resistance to Molten Vanadate-Sulfate Corrosion," *Surface and Coatings Tech.*, 39/40:89–96.

46 Namba, Y., Itoh, S., Okura, M., Takatsuka, K. and Kawata, S. "The Application of Plasma-Sprayed Coatings for Cooling Rolls in Continuous Annealing Lines," *Proc. Intl. Symp. Advanced Thermal Spraying Technology and Allied Coatings,* High Temp. Soc. Japan, Osaka, Japan, pp. 229–232, 1988.

47 James, A. S. and Matthews, A. "Thermal Stability of Partially-Yttria-Stabilized Zirconia Thermal Barrier Coatings Deposited by r.f. Plasma-Assisted Physical Vapour Deposition," *Surf. Coat. Technol.* 41:305–313, 1990.

48 Longo, F. M. "Coatings for Diesel Applications," *Proc. Intl. Symp. Advanced Thermal Spraying Technology and Allied Coatings,* High Temp. Soc. Japan, Osaka, Japan, pp. 125–130, 1988.

49 Henne, R., Bradke, M. V., Schiller, G., Schnurnberger, W. and Weber, W. "Vacuum Plasma Spraying of Oxide Electrocatalytic Materials," *Proc. 12th Int. Thermal Spraying Conf.*, London, 1989, I. A. Bucklow, ed., Abington Pub., Cambridge, 1:175–186, 1990.

CHAPTER 2

Crafting the Surface with Glow Discharge Plasmas

P. A. DEARNLEY[1]
T. BELL[2]
F. HOMBECK[3]

2.1 INTRODUCTION

IT has been some fifty years since the German entrepreneurial industrialist Bernhard Berghaus filed his original patents [1,2] on the plasma processing of metals. His work was profound in its originality. It provided the "foundation stone" for the two modern classes of *plasma surface engineering:*

- CLASS I: plasma nitriding, nitrocarburising and carburising—plasma-assisted *diffusion* methods
- CLASS II: sputter and evaporative source deposition—plasma-assisted *deposition* methods—the so-called Physical Vapour Deposition (PVD) methods

In this book there are a good number of chapters concerning aspects of the Class II plasma deposition processes. These are mainly used for the deposition of thin (≈ 2–4 μm) hard coatings of TiN for the wear protection of metal cutting and working tools. Class I processes are industrially more mature and somewhat more versatile—they *not only* confer wear resistance to engineering components but often improve rolling contact fatigue as well. For this reason, they are extensively used for the protection of steel gears.

Although this chapter is primarily concerned with Class I techniques, it

[1]Department of Chemical and Materials Engineering, University of Auckland, Private Bag, Auckland, New Zealand.
[2]School of Materials, University of Birmingham, UK.
[3]Klockner Ionon GmbH, Stauffenbergstrasse, Leverkusen, Germany.

is worthwhile at this juncture to summarise some of the more obvious distinctions between the two classes of plasma technology (Table 2.1).

From an academic standpoint, it is fascinating to note that both classes of treatment are frequently dependent upon the formation of "interstitial compounds" (i.e., transition metal nitrides and carbides) for their engineering success, although the mechanisms by which these compounds strengthen the surface vary considerably.

For Class I processes, interstitial elements (usually carbon or nitrogen) are conveyed to the surface of the component by the plasma. These subsequently become alloyed with the surface through the physical phenomenon of diffusion. For Class II processes, however, both interstitial elements and an appropriate transition metal (commonly titanium) are simultaneously transported to the surface by the plasma. These elements can combine *prior* or *subsequent to* deposition on the component surface to form a uniform deposit or coating. The essential mass transfer features of both classes of process and their interaction with the component (or "substrate") surface are schematically depicted in Figure 2.1.

Class II technologies have evolved principally for the wear protection of "hard" (>500 HV) tooling materials, such as cemented carbides and tool steels (e.g., hot and cold work die steels) [3]. This is no accident. Although such coatings have many admirable "ceramic" qualities, they are only several micrometers thick and, consequently, highly reliant upon the inherent strength of the component for mechanical support. There is little virtue in applying a ceramic coating of high hardness to an engineering

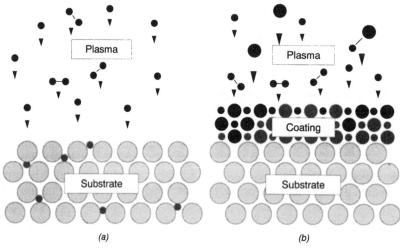

Figure 2.1 Schematic depiction of plasma surface engineering: (a) Class I diffusion processes; (b) Class II deposition processes.

TABLE 2.1. Overview of Class I and Class II Plasma Technologies.

Technology	Surface Engineering Principle	Temperature of Component during Treatment (°C)	Processing Pressure (mbar)	Typical Range of Attainable Properties	
				Max HV*	Depth (μm)
Plasma nitriding (Class I)	"Interstitial" solution of nitrogen	350–580	1.0–8.0	800–1200	50–1500
Plasma nitrocarburising (Class I)	Formation of ϵ-Fe$_{2-3}$N layer through "interstitial" solution of carbon and nitrogen	450–580	1.0–8.0	800–1100	5–10
Plasma carburising (Class I)	Solution of carbon, followed by quenching to form martensitic "case"	800–1000	5–20	700–800 (after temper)	50–2000
PVD (Class II)	Deposition of "interstitial" compound coatings via "reactive" nitrogen-bearing plasma	350–600	10^{-2}–10^{-3}	1800–2500 (TiN)	1–6

*Vickers microhardness (kg/mm^2).

23

component, only to find it "collapsing" into the relatively "soft" subsurface upon application of the working load.

Class I processes are mainly applied to ferrous alloys (i.e., steels and to some extent cast irons). Those that are plasma-nitrided on a regular industrial basis are listed in Table 2.2. Plasma nitriding develops strength up to 1.5 mm beneath the surface—some two orders of magnitude greater than the Class II deposition methods. Somewhat greater "case" depths are possible following plasma carburising (Table 2.1). Such hardened zones are an integral part of the component. It is for this reason that plasma-nitrided and carburised steels are able to sustain very high surface loadings.

2.2 PLASMA THERMOCHEMICAL DIFFUSION TREATMENTS

2.2.1 INTRODUCTION

Plasma thermochemical diffusion treatments are now widely accepted by the manufacturing sector in Europe, North America, and Japan. Increasingly, they have become the principal means by which gears, transmissions, and other machine parts have their wear resistance and fatigue endurance increased. Compared to conventional surface-hardening methods, plasma diffusion treatments have several significant advantages:

- reliability
- environmental cleanliness
- economy
- reduced cycle times
- flexibility
- easy masking
- excellent wear resistance
- minimal distortion
- anti-seizure properties
- superior control of microstructure

Although thermochemical diffusion treatments can be carried out using solid, liquid, or gaseous media, the technical, economic, and environmental constraints imposed by their use causes many heat treaters to change to plasma technology. Commercial enterprises are facing increased pressure from government legislation to "clean up" the environment and minimise industrial pollution. In this respect, the near mature technologies of plasma diffusion treatments are very attractive. They do not generally utilise toxic process gases and their total consumption of gas and electrical energy is modest compared to atmospheric gaseous methods.

TABLE 2.2. Typical Properties of Plasma-Nitrided Steels.

Type of Steels			Tensile Strength		Temperature Range		Surface Hardness	Nitriding Case Depth	
DIN	Material No.	AISI/SAE	(N/mm²)	KSI (1000 psi)	°C	°F	HVI	(mm)	(mil)
1) Structural Steels									
St 37–3	1.0116	~1020	370–450	55–65	550–580	1020–1080	200–350	0.3–0.8*	12–32
St 52–3	1.0570	~1019	550–700	80–100	550–580	1020–1080	200–450	0.3–0.8*	12–32
2) Carbon Steels									
C10	1.0301	1010	350–500	50–75	550–580	1020–1080	200–350	0.3–0.8*	12–32
C45	1.0503	1045	670–820	90–120	550–580	1020–1080	300–500	0.3–0.8*	12–32
3) Free Cutting Steels									
9S20K	1.0711	1212	350–550	50–80	550–580	1020–1080	200–400	0.3–0.8*	12–32
9SMnPb28	1.0718	12L13	400–550	60–80	550–580	1020–1080	200–400	0.3–0.8*	12–32
4) Cast Iron									
GG25	—	40B	≈250	≈35	530–570	980–1060	350–500	0.1	4
GGG60	—	80–55–06	≈600	≈85	530–570	980–1060	450–650	0.1–0.3	4–12
GTS55	—	—	≈550	≈80	520–560	970–1040	250–400	0.1	4
5) Powder Metal									
SINT	—	—	≥350	≥50	580	1080	260–350	0.1*	4
6) Cementation Steels									
16MnCr5	1.7131	5115	550–700	80–100	520–550	970–1020	500–700	0.3–0.8	12–32
15CrNi6	1.5919	≈4320	550–700	80–100	520–550	970–1020	500–650	0.03–0.6	12–24

(continued)

25

TABLE 2.2. (continued).

Type of Steels			Tensile Strength		Temperature Range		Surface Hardness	Nitriding Case Depth	
DIN	Material No.	AISI/SAE	(N/mm²)	KSI (1000 psi)	°C	°F	HVI	(mm)	(mil)
7) Spring Steels									
67SiCr5	1.7103	—	1500–1600	218–233	≤420	≤790	700–800	≤0.1	≤4
8) Heat Treatable Steels									
42CrMo4	1.7225	4140	800–1100	120–160	500–550	930–1020	550–650	0.3–0.5	12–20
30CrNiMo8	1.6580	≈4340	800–1100	120–160	490–540	910–1000	600–700	0.3–0.5	12–20
30CrMoV9	1.7707	—	800–1100	120–160	490–540	910–1000	750–850	0.2–0.5	8–20
9) Nitriding Steels									
31CrMol2	1.8515	EN40B	900–1200	130–175	490–540	910–1000	750–900	0.2–0.5	8–20
34CrAlNi7	1.8550	A355	850–1100	125–160	520–550	970–1020	900–1000	0.2–0.5	8–20
10) Heat Resisting Steels									
14CrMoV69	1.7735	≈514	900–1100	125–160	490–540	910–1000	750–900	0.4–0.8	16–32
56NiCrMoV7	1.2714	L16	1100–1300	160–185	450–550	840–1020	550–700	0.2–0.5	8–20
11) Hot Working Steels									
X32CrMoV33	1.2365	1110	1400–1600	200–235	480–530	890–990	900–1100	0.1–0.3	4–12
40CrMoV51	1.2344	1113	1400–1600	200–235	480–530	890–990	900–1100	0.1–0.3	4–12
12) Cold Working Steels									
X100CrMoV51	1.2363	A2	1800–2200	260–320	480–510	890–950	800–1000	0.1–0.2	4–8
X155CrVMo122	1.2379	D2	≥2400	≥350	480–510	890–950	900–1300	≈0.1	≈4

TABLE 2.2. (continued).

Type of Steels			Tensile Strength		Temperature Range		Surface Hardness	Nitriding Case Depth	
DIN	Material No.	AISI/SAE	(N/mm²)	KSI (1000 psi)	°C	°F	HVI	(mm)	(mil)
13) High Speed Steels									
S6-5-2	1.3343	M2	≥2400	≥350	480–510	890–950	1000–1250	0.02–0.1	0.8–4
S18-0-1	1.3355	T1	≥2400	≥350	480–510	890–950	1000–1250	0.02–0.1	0.8–4
14) Managing Steels									
X2NiCoMo18 85	1.6359	—	≥2100	≥305	460	860	850–950	0.05–0.1	2–4
15) Stainless Steels									
X20Cr13	1.4021	420	650–950	95–140	540–570	1000–1060	850–1050	0.1–0.2	4–8
X35CrMol7	1.4122	≈434	800–950	120–140	550–580	1020–1080	950–1100	0.1–0.2	4–8
X5CrNiI89	1.4301	304	500–750	75–110	550–580	1020–1080	900–1200	0.05–0.1	2–4
16) Valve Steels									
X45CrSi93	1.4718	HNV3	900–1100	130–160	530–560	990–1040	600–1000	0.02–0.1	0.8–4
X55CrMnNiN208	1.4875	—	900–1150	130–165	550–580	1020–1080	700–1000	0.02–0.1	0.8–4

2.2.2 CHARACTERISTICS OF PLASMA DIFFUSION METHODS

The Class I plasma technology described here was developed by Berghaus between 1935 and 1945. It was at this time that the first industrial applications were evaluated. From 1957 to 1967, two companies (Gesellschaft zur Forderung der Glimmentladungsforschung and Ionon—both of Cologne) collaborated in order to industrialise the early findings of Berghaus. From 1967 onwards, Klockner Ionon GmbH took on the industrial utilisation of the technology. They were the first company to actively transfer the technology into the manufacturing sector. Klockner Ionon GmbH is now the major world manufacturer of plasma-nitriding and carburising equipment. Furthermore, it has over twenty years of experience in operating its own commercial heat treatment facility, making it uniquely placed with regard to advising industrial clients on the commercial utilisation of the technology.

Before successful industrialisation of the process could take place, considerable development was required in the field of "power electronics." Perhaps the most crucial development was in the area of arc suppression control. In order to appreciate this development, it is necessary to understand more about the voltage-current characteristitics [4] of plasmas (Figure 2.2).

The most commonly utilised plasma in Class I and II processes is the so-called "anomalous" or "abnormal" DC glow discharge. It forms when a potential difference of a few hundred volts (or greater) is applied across a

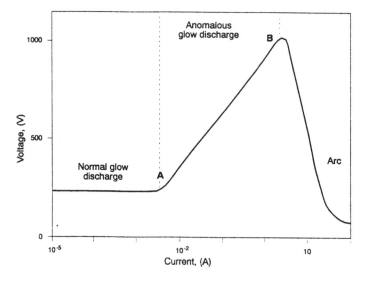

Figure 2.2 Typical voltage-current characteristics of a glow discharge plasma [4].

pair of electrodes placed in a chamber containing a gas (or mixture of gases) with a total pressure ranging from about 10^{-3} to 50 mbar. The glow discharge exploited in plasma surface engineering superficially looks like the "normal" glow discharge exploited in fluorescent lighting tubes. However, much higher power densities are possible. For Class I processes, these are typically ≈ 0.5–0.8 W/cm² (at pressures ≈ 1–8 mbar). For a given processing pressure, the voltage-current relationship has a positive and linear gradient (A-B in Figure 2.2).

If the current density is raised beyond a critical maximum value, the discharge rapidly decays or "collapses" into an arc discharge with detrimental consequences.

Hence, there is a requirement to "sense" an arc at the moment of its initiation and so prevent its propagation. This is achieved by rapidly, and momentarily, interrupting or "shutting down" the power supply. Sophisticated power supplies with built-in sensors and circuit breakers have progressively evolved over the years (Table 2.3). These are presently capable of sensing and shutting down an arc in 40 μs. It should be made clear that such devices are only used as a "safety net" in order to protect the production load from operator mistakes. With correct processing practice, the incidence of arc initiation is very rare.

Over the same period of history, the loading capability of plasma diffusion equipment has been steadily increased, and the related need for larger power supplies has been raised (Table 2.3). Perhaps one of the most impressive engineering feats has been the recent construction of a 1000-kW (1-MW) plasma-nitriding unit by Klockner Ionon GmbH for the URALMASH manufacturing plant in the former Soviet Union.

2.2.3 FEATURES OF A GLOW DISCHARGE PLASMA

2.2.3.1 Some Basic Concepts

In Class I processes, the components requiring treatment are made cathodic with respect to the surrounding chamber walls, which act as a counter electrode with positive ground potential (Figure 2.3). Electrical charge is then able to pass between the respective electrodes via the formation of a stable glow discharge plasma. Figure 2.4 shows a typical glow discharge plasma surrounding components during a plasma diffusion cycle.

Glow discharge plasmas markedly differ from low pressure gases. The molecules in a low pressure gas randomly move through space as depicted in Figure 2.5(a), whereas in a glow discharge, there is net drift ions and neutral molecules towards the cathode surface [Figure 2.5(b)]. A glow discharge is initiated by a primary ionisation event, possibly by the interac-

TABLE 2.3. Evolutionary Developments in Plasma-Processing Equipment (Courtesy of Klockner Ionon GmbH, Germany).

Year of Fabrication	1948–71	1971–74	1975–85	1986–90
Max power (kW)	72	150	450	1000
Total arc detection and shutdown time (μs)	65	60	40	40
Component charging weight (kg)	50–2000	1000–9000	up to 25,000	up to 30,000
Component heating mode	Ionic bombardment	Ionic bombardment	Cathodic heater and ionic bombardment	Cathodic heater, ionic bombardment and convective heater
Component cooling mode	In vacuum	In vacuum	In vacuum or gas quenching	In vacuum or gas quenching
Chamber type	Cold wall	Cold wall	Cold wall	Cold or hot wall

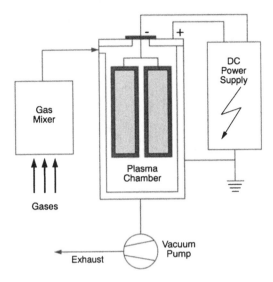

Figure 2.3 Schematic layout showing principal features of a commercial plasma-nitriding unit.

tion of a high energy cosmic ray with a neutral molecule or atom. Such effects happen continuously in nature. In the presence of a strong electric field, however, the ions are forced to move towards the cathode surface where they eventually strike or "bombard" it.

Following the initiation process, further ions are created by electron-atom or electron-molecule interactions. It is these latter processes that sustain the glow discharge. The electrons originate from the cathode surface

Figure 2.4 Glow discharge surrounding components during plasma nitriding.

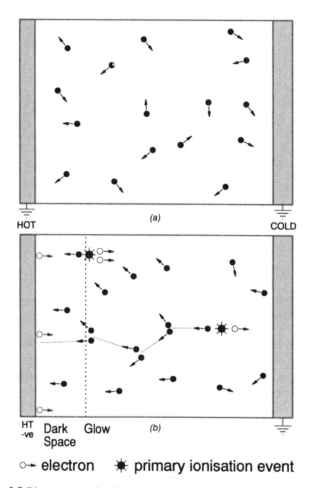

HOT   (a)   COLD

HT -ve   Dark   Glow   (b)
Space

○→ **electron**   ✳ **primary ionisation event**

Figure 2.5 Diagram comparing [4]: (a) gaseous treatment with (b) plasma treatment.

and are produced following the "ionic bombardment" of the cathode surface. Also, as the cathode becomes heated, electrons are emitted via thermionic emission.

2.2.3.2 Ionic Bombardment and Sputtering

At the pressures used in plasma diffusion treatments (1 to 10 mbar), the electrical field is at its strongest within a few millimeters of the cathodically charged surfaces. This means that the greatest voltage or potential drop takes place in this region—referred to as the "cathode fall" (Figure 2.6). Hence, ions formed from the process gas receive a massive acceleration across a comparatively short region of space. Following ionic bombardment, the incident ions give up some of their energy to the cathodically charged surfaces—in the form of heat and/or kinetic energy. The latter energy transfer is sufficiently large to cause the removal of some chemically bound surface atoms. The process of "knocking off" surface atoms by high energy ions is known as sputtering.

Sputtering is most efficient at high values of potential difference (voltage). In Class I processes, sputtering is utilised to greatest effect in removing oxide films formed on engineering component surfaces that might otherwise impede subsequent diffusion of the interstitial elements. This is especially important for the treatment of stainless steel and titanium components, which have a natural passivating surface film of metal oxide. Hence, high voltage values (≈ 800–1000 V) are often desirable at the commencement of a treatment cycle in order to impart "sputter cleaning." Following this initial period, moderate voltages are preferred (≈ 400–650 V), since excessive sputtering should be avoided.

2.2.4 PROCESS PRINCIPLES

2.2.4.1 Plasma Nitriding

There are many plasma diffusion treatments, but plasma nitriding (Klockner's "IONIT"®) remains the most widely practised. The related

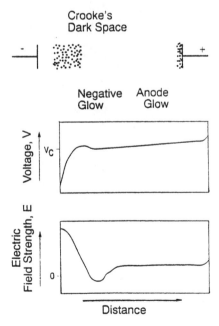

Figure 2.6 Spatial variation of electric field and potential in relation to visible (negative) glow.

processes of plasma nitrocarburising and plasma carburising have only become seriously utilised by industry within the past five years.

Plasma nitriding utilises hydrogen-nitrogen process gases. Some industrial companies prefer to use "cracked" NH_3 instead, but there is no technical necessity to do so.

Nitriding (nitrocarburising and carburising—see below) are diffusion-controlled processes. Therefore, the rate at which a given "case depth" develops is dependent on both temperature and time. For nitriding, temperatures range from 350 to 580°C, while times can vary between 5 and 100 hrs. The precise conditions selected depend upon the required case depth and the type of steel being treated. A great deal of skill is involved in selecting the right case depth for the right application. Naturally, there is always the economic reality to be met—getting the work processed in the most efficient manner. Typical nitriding temperature ranges, attainable case depths, and surface hardnesses for specific types of steel are compiled in Table 2.2.

Following nitriding, the structure of the case comprises two distinct zones (Figure 2.7): (1) an exterior compound layer and (2) an interior diffusion zone. The compound layer produced in ion nitriding is principally

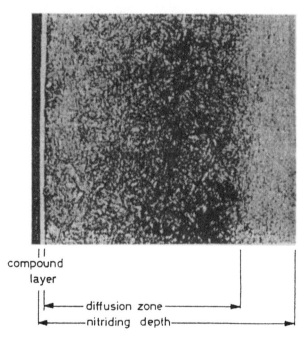

Figure 2.7 Microstructure of plasma-nitrided low alloy steel showing compound layer and adjacent diffusion zone [5].

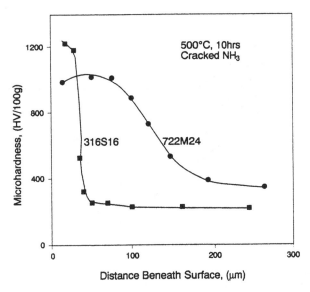

Figure 2.8 Typical microhardness profiles obtained after plasma nitriding low alloy (722M24) and high alloy (316S16) steels.

γ'-Fe$_4$N [5]. It varies in thickness, but does usually exceed ≈ 5 μm, unlike gaseous nitriding where very thick (≈ 50 μm) compound layers can be produced. In situations where even 5 μm of γ'-Fe$_4$N is regarded as detrimental, it is possible to reduce this to below 1 μm (by appropriate control of the nitrogen content of the plasma).

The interior diffusion zone constitutes the principal feature of a nitrided "case." It is the region in which the greatest hardness is observed. A typical microhardness profile, through a nitrided case of a low alloy steel, is shown in Figure 2.8. By way of comparison, a hardness profile for a high alloy (stainless steel) is also given, which for the same treatment conditions is much shallower.

In order to achieve the hardening effect within the diffusion zone, it is necessary that the steel contain specific nitride-forming elements capable of forming submicroscopic precipitates. The most common of these are Cr, Al, V, and Ti. The "alloy nitride" precipitates (like CrN) formed during nitriding are extremely small, perhaps only a few hundred atomic planes in magnitude. They do not form a definite boundary with the steel matrix, i.e., they are coherent with it. The precipitates are very effective in strengthening steel because they make plastic deformation an extremely difficult process, i.e., the usual mechanisms of internal slip are impeded. The higher the alloying element content (e.g., Cr), the stronger the interaction with the nitrogen and the shallower the nitrided case (for a given

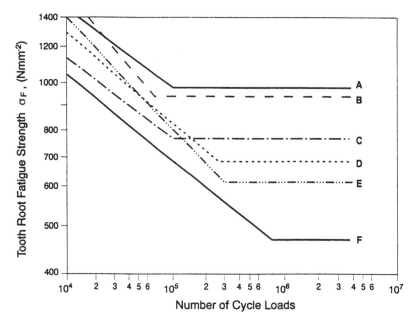

Figure 2.9 The "tooth" root fatigue strength of plasma-nitrided and -nitrocarburised steel gear wheels. ND = depth of diffusion zone [6].

treatment time). This can be seen in Figure 2.8 where the more highly alloyed (316S16) stainless steel exhibits a shallower nitrided case than that attained by the conventional low alloy (722M24) nitriding steel.

A practical consequence of the formation of CrN precipitates is that they try to "expand" the surrounding steel matrix. This results in the diffusion zone being placed in a state of residual compressive elastic stress (or more precisely—residual elastic strain). This has a very positive engineering consequence: any crack trying to propagate through this zone will experience a stress state opposing crack growth. It is for this reason that there is a marked increase in rotation bending fatigue endurance (and fatigue strength) following plasma nitriding. This factor increases the rolling contact fatigue resistance of gears [6], as shown in Figure 2.9.

2.2.4.2 Plasma Nitrocarburising

Plain carbon steels do not respond to nitriding but can be satisfactorily plasma nitrocarburised. Plasma nitrocarburising, like nitriding, produces both an exterior compound layer and an internal diffusion zone. However, as it is not possible to strengthen plain carbon steel matrices by appropriate nitride precipitation, the technique is aimed at stimulating the growth of the exterior iron nitride layer. In particular, it is the ϵ-Fe$_{2-3}$N phase that is sought. This phase is thought to be especially suited to conveying anti-galling behaviour, while at the same time providing enhanced sliding wear resistance [7]. Hence, the objective in nitrocarburising is to produce a relatively thick (≈ 5–10 μm) continuous layer of ϵ-Fe$_{2-3}$N, resembling a coating in appearance, although it forms via a diffusional interaction with the steel substrate. It is *not* deposited from the plasma.

In plasma nitrocarburising, a carbon-bearing gas (usually methane) is added to the normal nitrogen-hydrogen plasma. The metallurgical function of the carbon is to "stabilise" the formation of the ϵ-Fe$_{2-3}$N phase, as a cursory examination of the Fe-C-N phase diagram will confirm. If this is not done, γ'-Fe$_4$N (the other nitride of iron) or a mixture of the two nitrides will form instead.

It would appear that a certain amount of skill is required in order to optimise the plasma-nitrocarburising process. Since carbon is an important factor in the stabilisation of the ϵ-Fe$_{2-3}$N phase, the contribution of carbon from the substrate is important. The addition of *minor* quantities of an oxygen-bearing gas to an established CH$_4$/H$_2$/N$_2$ plasma can also be helpful. This can promote the formation of ϵ-Fe$_{2-3}$N [8].

Recently, a special development project was set up by Klockner Ionon GmbH in order to devise a processing route suitable for the nitrocarburising of powder metallurgy parts. Such components contain a high-volume fraction of micro-pores. During conventional salt bath nitriding, salt enters the pores and cannot be subsequently removed. This results in degradation of the component through internal corrosion. The development project demonstrated *both* the viability and advantages of the plasma-nitrocarburising method. The corrosion problems were obviated. Hence, nowadays, up to 4500 powder metallurgy chain gears can be routinely plasma-nitrocarburised in one production cycle. Treatment times are deliberately kept short (≈ 2–4 hrs). This is sufficient to develop a surface layer of ϵ-Fe$_{2-3}$N that is ≈ 6-μm thick, while at the same time avoiding the formation of embrittling subsurface iron nitrides—a feature of gaseous nitrided materials. Figure 2.9 shows that the fatigue strength of the tooth root of plasma-nitrocarburised gears (curves E and F) are inferior to those developed in nitrided gears (curves A, B, C and D). For this reason, nitrocarburised gears are used in less demanding applications than nitrided gears.

2.2.4.3 Plasma Carburising

Plasma carburising has only been established as an industrial scale process within the past five years, even though its feasibility was known for a considerable time [9,10]. Unlike plasma nitriding and nitrocarburising, plasma carburising is an austenitic heat treatment. This means that the appropriate grade of steel must be heated into the austenitic phase region (typically 850–1000°C) before it can be carburised. Following carburising, the components must be rapidly cooled via a sealed oil quench or (if appropriate) a high pressure gas quench. This is followed by the usual low temperature temper.

If components are to be heated above 800°C by a glow discharge plasma, very high plasma power densities (≈ 7 W/cm²) are required [11]. These levels of power density can be unacceptable because they lie close to the arc transition region. This problem is most easily obviated by the use of radiant heaters located within the plasma chamber. This allows for a safer working process while at the same time assuring a more acceptable carbon potential. High plasma power densities equate to a high carbon potential. In the extreme, this leads to the formation of soot. In fact, it is necessary to use $\approx 1.5\%$ CH_4 in H_2 to avoid this effect [11]. The use of radiant heaters, however, allows more moderate plasma power densities (≈ 1 W/cm²) to be used. This means that the carbon potential (for a given gas composition) is actually reduced—so much so that it becomes possible to use 100% CH_4 without any soot formation [10]. In any event, the mass transfer in plasma carburising is considered to be very high, and so it is normal practice to adopt the "boost-diffuse" method [4]. This is the most efficient way of transferring surface carbon into the "case." Typical carbon concentration and microhardness profiles resulting from this approach are given in Figure 2.10. Hence, in certain respects, a modern plasma-carburising unit resembles a traditional vacuum furnace, except there is no requirement to use a diffusion pump. One industrial example is shown in Figure 2.11.

2.2.5 OPERATIONAL FEATURES

A modern plasma-nitriding facility is shown in Figure 2.12, while a schematic depiction of the process is given in Figure 2.3. All plasma nitriding equipment manufactured by Klockner Ionon GmbH comprises:

- vacuum chamber(s)
- vacuum system
- gas supply, distribution, and mixing system
- power supply
- microprocessor control

2.2.5.1 Vacuum Chambers

Vacuum chambers can be manufactured in a number of configurations including "horizontal," "pit," "bell," and "combination" units. Specific

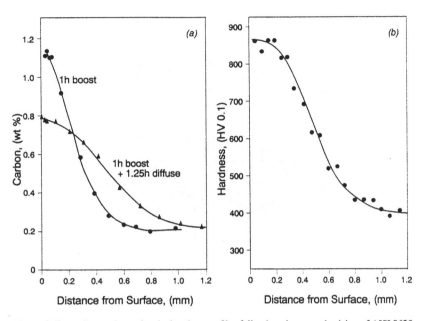

Figure 2.10 (a) Comparison of typical carbon profiles following plasma carburising of AISI 8620 with and without subsequent diffusion treatment; and (b) microhardness of "boost-diffused" AISI 8620 steel sample after oil quenching [4].

Figure 2.11 Small industrial plasma-carburising unit built by Klockner Ionon GmbH, Germany.

Figure 2.12 Large dual chamber plasma-nitriding unit built by Klockner Ionon GmbH, Germany.

chamber dimensions are determined by the workpiece dimensions and the production loading requirements. Modern chambers have a broad range of capability, depending upon the needs of the customer. This might range from a few kilograms in one case to loads of 30 tonnes in another. Flexibility is also important. One user of Klockner equipment needs to nitride 70,000, 1-mm diameter balls (for ballpoint pens) during one cycle. At other times, the unit can be reconfigured to nitride a 26-tonne trunnion bearing (for an eccentric press).

2.2.5.2 Power Supply Specification

At the design stage, it is necessary to consider the "energy balance" for a given plasma system. A compromise has to be made between insulation and heat loss (cooling) on the one hand and the thermal energy supplied by the plasma on the other. From this point of view the *range* of possible workpiece surface areas should be known, since these values determine the range of plasma power density that the workpieces receive.

This range of power density *must* exceed a *critical minimum value*, in order for satisfactory hardening to be achieved, i.e., power density is a function of chemical potential. This effect is illustrated for plasma nitriding in Figure 2.13. When the power density falls below 0.4 W/cm² for 42CrMo4 steel, the surface hardness falls below 600 HV. This is unsatisfactory, i.e., the nitrogen potential is too low [12].

The electrical energy required for establishing a glow discharge plasma, can be of two types: (1) pulsing and (2) pulse-pause. Klockner Ionon can supply both. Regardless of the type used, the power density (integrated with respect to time) must exceed the same critical value (for a given type of workpiece material). This is clearly seen in Figure 2.13. This means that the *frequency* of the electrical pulses does not intrinsically influence the chemical potential at the surface. The important factor is the integrated power density (W·hr/cm²).

Some interesting statistical comparisons have been made comparing the variation in surface hardnesses achieved after nitriding with the two types of power supply for a large number of nitrided components. These results are summarised in Figure 2.14. Unit A, manufactured by Klockner Ionon, was a cold wall (water-cooled) plasma-nitriding unit with a pulsing power supply, while Unit B was a hot wall (insulated) system with pulse-pause power supply, manufactured by a competitor of Klockner. The results (Figure 2.14) show a much wider scatter in surface hardness for Unit B than for Unit A, i.e., the Gaussian distribution was more tightly scattered around the desired mean surface hardness value of 650 HV for Unit A than

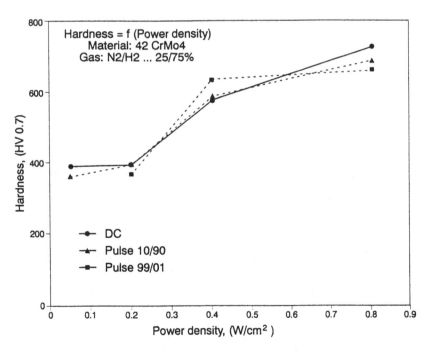

Figure 2.13 Relationship between power density and surface hardness for plasma-nitrided 42CrMo4 steel. There is no significant difference in the relationship, whether using pulsing (DC) or pulse-pause power variants [12].

Quality Control According to Ford's Reliability Methods

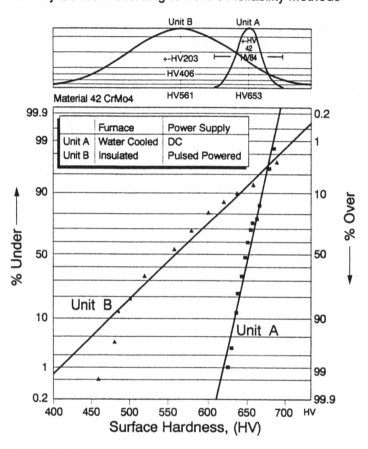

Figure 2.14 Surface hardness distributions for components plasma-nitrided with: "Unit A"—a cold walled system equipped with pulsing power and "Unit B"—a hot walled system equipped with pulse-pause power [12].

for Unit B. The mean surface hardness for the components produced by Unit B (561 HV) was well below the desired control value.

The probable reason for the disparity is that the plasma power density used by Unit B was too low, i.e., there was too much radiant heating and/or too much thermal insulation. There is therefore a practical requirement to assure efficient heat transfer from the load to the chamber walls, otherwise the plasma power density that can be permitted will be too small to assure homogeneous nitriding.

2.2.5.3 Process Control

The modern plasma-nitriding and -carburising units constructed by Klockner Ionon GmbH utilise a microprocessor system that automatically regulates all the treatment parameters, from initial evacuation to final cooling. The microprocessor can store up to twenty-four different processing programs. During a given production cycle, it is possible to interrupt and "edit" the program any time. The microprocessor can also be interfaced to a host computer, possibly located outside the production shop area, thus allowing remote monitoring. This is particularly useful when running more than one plasma unit.

Temperature control is assured by using up to four independent thermocouples strategically located within the processing load [13]. It is always the "hottest" thermocouple that is used to control the temperature of the load. Temperature uniformity is always within $\pm 10°C$.

The gas mixing device (for up to four different gases) achieves mixtures that are within ± 1.0 vol% of the desired values. When only small quantities of a particular gas are required, greater accuracy (within ± 0.1 vol%) is possible.

2.2.5.4 Reduced Treatment Times

This effect is principally seen when comparing plasma processes with gaseous treatments. In contrast to gaseous thermochemical treatments, the strong electric field in plasma treatments (Figure 2.6) is responsible for attracting a large flux of ions and energetic neutrals towards the cathodic workpiece surface [Figure 2.1(a)]. Hence, far greater numbers of "active" species will arrive at the component surface, per unit time, for a glow discharge environment than for a gaseous system of equivalent pressure.

Passive oxide films are always present on engineering surfaces. These are rapidly removed by sputtering, a phenomenon that takes place throughout the plasma heat treatment cycle. Conventional thermochemical processes cannot depassivate surfaces in the same way. Depassivation through sputtering therefore assures a rapid mass transfer of interstitial elements from the plasma to the subsurface of the component.

One striking example of the difference between subatmospheric gaseous (vacuum) and plasma carburising is shown in Figure 2.15. A much higher total carbon (area under curve) was introduced into the carburised case by the plasma process, using a methane partial pressure of 0.4 mbar, than was possible by using 130 or 260 mbar of the same gas under subatmospheric conditions.

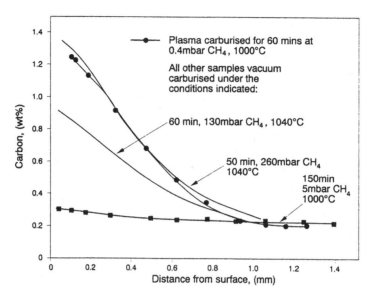

Figure 2.15 Comparison of carbon profiles produced by plasma and low pressure gaseous carburising [4].

2.2.5.5 Reduced Energy Consumption

For cold-walled plasma units, only the cathodically charged components are heated, while for traditional hot wall gaseous systems, the walls *and* the work load are radiantly heated. Marciniak and Karpinski [14] have undertaken a detailed energy assessment comparing the two types of unit with respect to nitriding. They found that when both types of unit were fully loaded, there was little difference in the total consumption of electricity. In the case of *partially* loaded units (a not uncommon industrial situation), the plasma-nitriding method was the most economic.

However, these workers have ignored the positive effect of depassivation of the metal surface, which is of most significance over short processing cycles or when nitriding "difficult" materials such as stainless steels. Furthermore, the total consumption of process gases is reduced greatly for plasma systems compared to gaseous units. The latter two aspects provide plasma systems with lower running costs, while at the same time producing comparatively minimal industrial pollution. An equivalent analysis of hot walled plasma systems has yet to be made.

2.2.5.6 Environmental Aspects

Plasma-nitriding, -nitrocarburising, and -carburising treatments produce no undesirable or toxic waste products; there are no risks of explosion and there are no significant noise emissions.

2.2.6 FUTURE PLASMA DIFFUSION TREATMENTS

2.2.6.1 Plasma Nitriding Titanium

In recent years, the application of titanium alloys has been increasing. This has inevitably meant that titanium alloys are being "pushed to their limit" and used in circumstances where *wear* severely limits their performance. There has been some considerable effort to improve the wear resistance of titanium alloys by plasma nitriding. Unlike steels, fairly high nitriding temperatures are necessary ($>700\,°C$) in order to produce even a moderate "case" depth. This is because the kinetics of nitrogen diffusion in titanium are orders of magnitude slower than in ferrous workpieces. There is also the added disadvantage that there is no precipitation strengthening mechanism of the type described above for alloy steels. Instead, strengthening of the titanium is only achieved through interstitial solid solution hardening. Fortunately, a duplex surface layer of TiN/Ti_2N also forms via the usual diffusion "driven" nucleation and growth processes. It is this layer (Figure 2.16) that is responsible for providing tribological protection to titanium components. The thickness of the layer is dependent on temperature and time. It is usually not practicable to achieve a layer that is greater than 10 μm.

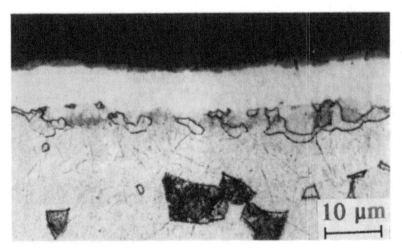

Figure 2.16 Microsection of plasma-nitrided Ti-6Al-4V showing and external layer of TiN (dark grey) and an adjacent layer of Ti_2N (light grey). The latter phase is approximately five times thicker [16].

2.2.6.2 Duplex Treatments

There might be some advantage to be gained from combining plasma nitriding with an appropriate PVD treatment, so that the advantages of a ceramic coating like TiN can be exploited for low alloy steels. The "idea" is to obviate the problem of inadequate substrate support (discussed in Section 2.1) by strengthening it with a prior nitriding treatment. Preliminary investigations have shown this to be a viable possibility [15].

Another interesting possibility is to nitride the substrate *after* covering the surface with a thin hard ceramic layer [16]. This method is dependent on the diffusivity of nitrogen through a given ceramic film. This may be quite difficult in the case of TiN. However, it is definitely possible to diffuse nitrogen through an iron boride layer (developed by a prior austenitic boriding treatment). One such result is shown in Figure 2.17. The subsequent nitriding operation has successfully raised the hardness of the substrate to a value (≈ 800 HV) that ought to provide adequate support for the very hard (1900 HV) iron boride layer. This kind of duplex treatment

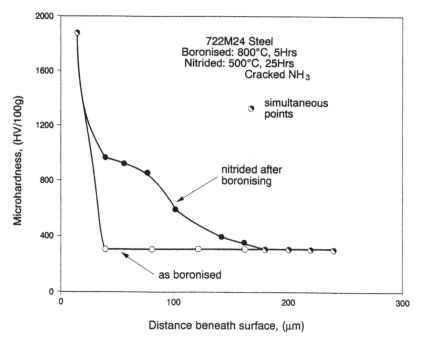

Figure 2.17 Comparison of microhardness profile obtained for boronised low alloy steel. The strength of the steel substrate can be markedly improved by carrying out a subsequent plasma-nitriding operation [16].

could be entirely carried out within a conventional plasma diffusion unit, providing a few adaptations were made.

2.3 SUMMARY

Plasma diffusion treatments are amongst the most useful and well-established industrial scale plasma methods that are available to the surface engineer for *crafting* the surface. Of these, plasma nitriding remains the most widely practiced.

Significant development steps have taken place over the past five years in the related areas of plasma nitrocarburising and carburising and their industrial credibility proven. These technologies are being progressively introduced into large-scale industrial manufacturing facilities; they are more efficient, have less "down-time," and cause less industrial pollution than equivalent conventional thermochemical diffusion processes.

2.4 REFERENCES

1 Berghaus, B. German Patent, DPR 668,639, 1932.
2 Berghaus, B. German Patent, DPR 851,560, 1939.
3 Dearnley, P. A. and Bell, T. *Ceram. Eng. Sci. Proc.*, 9(9–10):1137–1158, 1988.
4 Booth, M., Farrell, T. and Johnson, R. H. *Heat Treat. Met.* (2):45–52, 1983.
5 Edenhofer, B. *Heat Treat. Met.* (2):59–67, 1974.
6 Weck, M. and Schlötermann, K. *Metallurgia*, (8):328–332, 1984.
7 Bell, T. *Heat Treat. Met.* (2):39–49, 1975.
8 Hadfield, J. B. M. Sc. Thesis, University of Birmingham, UK, 1986.
9 Edenhofer, B. *Hart-Tech Mitt.*, 29(3):165–170, 1973.
10 Grube, W. L. and Gay, J. G. *Met. Trans.*, A, 9:165–172, 1978.
11 Edenhofer, B. *Proc. Conf. on Plasma Surface Engineering, Garmisch-Partenkirchen*, 1:257–268, 1988.
12 Hombeck, F., Oppel, W. and Remges, W. *Proc. Conf on Plasma Surface Engineering*, Garmisch-Partenkirchen, 1:257–268, 1988.
13 Remges, W. and Oppel, W. *Wärm. Int.*, 33(6/7):349–353, 1984.
14 Marciniak, A. and Karpinski, T. *Ind. Heat.*, 47(4):42–44, 1980.
15 Sun, U. Ph.D. Thesis, University of Birmingham, UK, 1989.
16 Dearnley, P. A. Unpublished research, 1988.

CHAPTER 3

Surface Engineering of Polymers for Biomedical Applications

H. J. GRIESSER[1,*]
R. C. CHATELIER[1]
T. R. GENGENBACH[1]
Z. R. VASIC[1]
G. JOHNSON[2]
J. G. STEELE[2]

3.1 INTRODUCTION

FOR the designed optimization of the performance of polymers in many modern applications, the ability to fabricate specific surface compositions tailored to the particular application has become an important consideration. The need for designed, well-controlled chemical compositions of polymer surfaces arises from the fact that interfacial phenomena define properties that are crucial to the service performance of a particular device. Examples of applications where polymer surface properties are important include wetting, printing, biomedical devices, and membranes. In all cases, molecules from the "environment" approach the polymer surface (or the surface inside pores leading through the polymer) and experience interfacial forces due to electrostatic and electron cloud interactions. These interfacial forces act across extremely short distances; therefore, only the chemical groups located within a few Angstroms of the surface of the polymeric material exert a significant influence across the interface towards the environment. It thus becomes possible, in principle, to design guided approaches towards device optimization by altering and controlling interfacial interactions of polymers by the provision of appropriate chemical structures in the surface layers.

*Author for correspondence.

[1]Division of Chemicals and Polymers, CSIRO, Private Bag 10, Clayton, Victoria 3168, Australia.
[2]Division of Biomolecular Engineering, CSIRO, P.O. Box 184, North Ryde, NSW 2113, Australia.

Surface modification techniques have attracted increasing interest for the fabrication of specialty polymers from commodity bulk polymers. An important reason is that the surfaces of bulk synthesized polymers generally are mobile and unpredictable. The synthesis of novel specialty polymers often is not only expensive, but usually also provides little control over the composition of the surface on account of chain movements, segregation, and the dependence of surface composition on fabrication conditions. Often, the chemical composition of a polymer surface is not identical to and predictable from the composition of the bulk polymer [1]. For example, microdomain segregation takes place in polyetherurethanes, producing a surface layer that contains mainly or exclusively polyether segments [2]. The polymer surface will adopt a chemical composition that allows it to minimize the interfacial energy. When in contact with air, hydrophobic groups are favoured at the interface. However, this process does not stop once the polymer has solidified; polymer chains continue to display a certain mobility in the solid state, and these thermally driven motions of the polymer chains will continue to allow the gradual preferential accumulation at the surface of those chemical groups that best minimize interfacial energy [1]. Accordingly, on immersion in polar liquids, hydrophilic groups will emerge from the bulk and accumulate at the surface. As a result, polymer surface compositions can be unpredictable, as well as varying with time, as they are driven to respond to changes in their environment.

In biomedical applications, further complications arise. When a polymer surface is placed in contact with a biological environment (for instance, blood, body fluids, tissue), adsorption of proteins onto the polymer surface occurs immediately. However, proteins consist of polar and nonpolar segments and, like polymers, can reorient their structure in response to forces that act on the constituent atoms [3]. Hence, when a protein approaches a polymer surface, it is likely that a process of mutual rearrangements of structures occurs until an energy minimum is reached. This can lead to structural rearrangements of the protein, which are so extensive as to render it incapable of performing its original biological function; the protein is said to have become denatured. In addition, biological media contain a large number of mobile molecules, among them various proteins, all of which can competitively adsorb on the polymer surface, but only a few proteins will provide a beneficial function such as the promotion of cell attachment and growth while others will block or interfere. The aim of designed biocompatibility studies could thus be the fabrication of polymer surfaces that selectively attract and adsorb from complex mixtures those proteins that confer a desired biological response. However, at present the details of the protein adsorption and denaturation processes are poorly understood, and the analysis and control of the relevant interfacial forces remain a considerable challenge.

The study of interactions between polymers and biological media requires the fabrication of well-characterized and stable polymer surfaces upon which protein adsorption and cell growth experiments can be performed and correlated with surface composition, so that, ultimately, the role of various chemical groups can be clarified and models formulated for the interfacial interactions. To date, however, many polymers used or intended for biomedical applications have not been surface designed and analyzed in detail, and much biomedical testing is not supported by sufficient surface analytical data. Thus, at present, the most promising strategy appears to be the fabrication and analysis of a range of polymer surface compositions and their evaluation in model biological tests, with the aim of identifying the relevant surface groups, their most effective density, and their interaction with protein mixtures. Fine tuning can then be done for specific applications.

An attractive, potentially cost-effective way of fabricating polymers with specific surface compositions is to attach the desired groups by surface modification technologies after fabrication of the bulk polymer material, which is chosen on the basis of desired mechanical properties. Surface modification techniques have been used for many years but often in ways that are chemically poorly understood. The techniques of flame treatment, acid etching, and corona discharge treatment, which have been used extensively in industrial applications, produce a variety of new polar surfaces [4]. This is quite acceptable in wetting and printing, but for applications of polymers in biomedical devices, the presence of a polar surface is, according to our data to be presented, not sufficient. Metabolic constituents and proteins appear to interact very specifically and selectively with particular chemical groups on the polymer surface, and possibly the density of such groups is also important. Therefore, for more sophisticated applications such as biomedical, the conventional, chemically ill-defined, and unspecific surface treatments appear less suitable for the development of designed polymer surfaces. Furthermore, corona, acid, and flame treatments appear to be nonuniform on a microscopic level and produce some etching; and surface topography, too, can alter the response of cells, interfering with the effects of surface chemistry.

Low pressure gas plasma techniques [5–9] present advantages for the surface modification of commodity polymer substrates because an extended range of surface chemistries can be obtained with each individual process gas by modification of the experimental variables. While the intensity of the plasma at the surface is generally strong, the penetration depth of the treatment is even and very low at a reaction level sufficient for useful surface modification. Thus, bulk polymer properties are altered to a lesser extent in a plasma than with alternative treatment technologies.

Plasma surface treatment and plasma polymerization have been explored in order to produce novel polymer surfaces and have evaluated their performance in model biological tests involving the attachment and growth of human cell lines on the polymer surface in a culture dish. By correlating surface analytical data and biological performance with experimental plasma conditions, the ultimate aim is to be able to "design" biocompatible surfaces that are stable and contain a controllable density of specific chemical groups and to understand the biochemical basis for their improved performance. To support such studies, it is also necessary to develop refined surface analysis techniques.

3.2 PLASMA SURFACE MODIFICATION TECHNIQUES

The composition of gas plasmas is exceedingly complex, even when starting from simple molecules such as oxygen, because of the diversity of the processes following the initial ionization of a molecule in the gas phase. Radicals and ions produced can rearrange, fragment, or react with other species; hence, when organic molecules are fed into a gas plasma, a myriad of reactions and products can be produced. Typically, *inorganic* molecules induce reactions on surfaces in contact with the plasma and lead to the formation of new groups to a very shallow depth without significant removal or deposition of material although etching often does occur on prolonged exposure. This is called plasma surface treatment. In contrast, *organic* molecules generally form new carbon-carbon bonds in the plasma and ultimately condense on surfaces in the vicinity in the form of thin polymeric film coatings—a process called plasma polymerization [5–9]. Plasma surface treatment of conventional polymers and plasma polymerization make use of low temperature plasmas, where ions typically have low kinetic energy, whereas electrons can have energies up to 20 eV, although the average is considerably lower. Low-temperature plasmas that are stable with time and of technological importance are typically operated at low pressures of 1 torr or less [5].

Plasma surface modification is related to corona discharge treatment; coronas are usually established at atmospheric pressure and contain time-dependent, nonuniform areas of gas phase ionization, thus producing arcs and nonuniformity of the treatment on a micrometre scale, whereas low pressure plasmas can be made uniform and stable. For the surface treatment of insulating materials such as plastics, alternating electric fields are generally used to drive the corona or plasma discharge. Thus, the surface of a polymeric substrate exposed to the plasma is bombarded alternatingly with electrons and negative ions, and positive ions.

The effective depth of plasma treatment in nondepositing gases is very low; the chemical composition of the polymer is modified only to a depth of a few nm [10] because the impinging ions do not possess sufficient energy for deeper penetration into the polymer. Thus, changes to the physical properties of the treated material (substrate) are minimal, even under conditions that markedly alter the composition of the surface layers. Likewise, the deposition of a thin film coating often does not measurably alter the mechanical properties of the resulting composite compared with the untreated substrate. Therefore, plasma processes allow, in principle, the optimization of surface properties independent of bulk properties. The much lower penetration depth of plasma treatment and the ultrathin coatings feasible by plasma polymerization, compared with alternative technologies, make plasma techniques the methods of choice where it is important to modify the surface properties without significantly affecting the bulk.

Plasma polymerization creates a polymeric layer directly from vapour phase material; coating thickness can be controlled easily and on-line. Plasma polymer films are usually very smooth and uniform due to the molecular nature of the film-building process; also, because the coating is applied directly from the gas phase, the plasma process avoids problems of wetting and nonuniform spreading, which can occur in conventional solvent coating processes. Full surface coverage can be obtained at a coated thickness of a few nanometres. Adhesion to substrates is excellent, in general, because free electrons, which travel much faster, impinge on the substrate surface and break some bonds prior to coating, thus creating sites for covalent attachment of the plasma coating. "Monomer" activation by homolytic bond scission or ionization and electron-activated processes in the plasma atmosphere obviate the need for reactive monomers, their stabilization, and the use of chemical initiators. Simple compounds such as hexane and benzene are useful "monomers" and readily form plasma polymer coatings.

While plasmas are versatile in being able to activate normally unreactive chemical structures, produce novel chemistries, and polymerize most organic compounds that can be evaporated, the drawback is that this plasma activation also allows a large number of possible reactions to occur in a plasma; these processes are not well understood. Low-temperature plasmas are nonthermal; classical chemical principles do not apply. Modified surfaces and plasma polymers are not formed by mechanisms analogous to those of conventional chemical reactions and polymerizations, respectively. Plasma polymers do not consist of regular repeat units built up from monomers by sequential additions—they are thus not polymers in the traditional sense—but, instead, consist of random structures that incorporate various fragments derived from the "monomer" initially supplied. Bond breakage by electron impact is usually not very selective; any one of several bonds can have a finite probability of scission.

It also appears that multiple bond scission in a given monomer molecule occurs with considerable frequency, as evidenced by the observation that plasma polymers often are extensively crosslinked in random, three-dimensional ways; this makes them insoluble in all solvents and of essentially infinite molecular weight. Therefore, at present plasma polymerization cannot match conventional polymerization processes for the degree of control of chemical composition of the polymers produced. The relationship between the experimental plasma parameters (pressure, gas flow rate, electrical power, frequency of the applied field, geometry, and others) and the composition of the films produced is very complex and poorly understood. A continuing challenge is the elucidation of parameters, which, for a particular gas, provide the highest relative yield of a desired pathway or chemical group on the surface. In summary, the fabrication of fully "designed" surfaces by plasma techniques is not yet feasible.

Plasma techniques are applicable to all polymers and especially useful for surface modification of polymers that are difficult to treat by conventional chemical methods, such as fluorocarbons [11–13]. Fluorocarbons are particularly unsuitable for biomedical applications, which demand attachment of cells, since cells appear incapable of growing on these strongly hydrophobic surfaces. However, the bulk mechanical properties of fluoropolymers often are outstanding, making them excellent candidates for surface modification. Poly(tetrafluoroethylene) (Teflon, PTFE) and fluorinated ethylene propylene copolymer (FEP) have been chosen as model polymer substrates and samples comprising thin metal coatings on polyimide (Kapton) as model inorganic substrates. The Co_3Cr coating was fabricated by sputter deposition and the Al coating by electron beam evaporation. The fluorocarbon polymers were commercial samples, while the metallized Kapton tapes were kindly provided by C. Brucker and R. Spahn of the Research Laboratories, Eastman Kodak Company, Rochester, New York, U.S.A.

3.3 EXPERIMENTAL

The reactor and ancillary equipment used for plasma treatment and plasma polymerization experiments have been described previously [14]. The reactor differs substantially from conventional bell jar systems in that the monomer is confined in a flow channel until it reaches the plasma zone, thus allowing accurate determination of the amount of gas provided to the plasma discharge and efficient utilization plasma conditions onto moving substrate (1/2-inch wide tape). In this way, the effects of spatial variations of the deposition rate are eliminated, and uniform coatings can be produced onto several metres of tape, which moves at speeds of 0.05 to

6 m/min. Plasma generators with the frequencies 125 to 375 kHz (ENI HPG-2), 700 kHz (custom built), and 13.56 MHz (ENI ACG-3) were used.

PTFE was only available in sheet form and was treated in the stationary mode by attaching samples to the face of the plasma electrodes by thin double-sized sticky tape. Some FEP samples were also treated in the stationary mode. FEP and Co_3Cr/Kapton tapes were treated in the moving substrate mode by transporting extended lengths of tape through the plasma region following stabilization of the plasma.

PTFE and FEP samples were plasma surface–modified using oxygen, water vapour, argon, and air as process gases, with the samples attached to the face of the elctrodes. Oxygen and argon (high purity grade) were provided from cylinders, air was taken from the laboratory (low humidity), and water vapour was obtained from the boiloff at room temperature of distilled water placed in a round bottom flask and evacuated. Except for water vapour, whose flow was manually controlled and measured by the rate of pressure increase on closing the valve to the pump, the flow rate of the process gas was measured by commercial gas flow meters (MKS) and was typically in the range of 4 to 15 sccm/min.

Plasma polymer films were deposited to various thickness in the range of ~50 to 150 nm onto moving substrate tapes under a stabilized plasma condition. The metallized Co_3Cr/Kapton tape enabled on-line observation and control of the coating quality, assessment of the coating thickness by the interference colouring, and IR spectroscopic analysis of the plasma polymer coatings without interference from substrate bands. The monomer flow rate had to be set and determined manually because the commercial mass flow controllers imposed a pressure drop, which in most cases, exceeded the room temperature vapour pressure of the monomer liquid; it was determined prior to deposition by measuring the rate of pressure increase in the reactor with the exit valve closed. The monomer liquids used were all of the best available commercial grade. Any more volatile impurities that may have been present were removed by pumping vapour off the monomer liquid for a few minutes prior to plasma experimentation.

Contact angles with distilled water were measured using a modified Kernco G-II Contact Angle Meter equipped with a syringe incorporating a reversible plunger driven by a micrometer to allow determination of advancing, sessile, and receding angles. The stability of the contact angles of selected plasma polymer samples was assessed on another custom-built contact angle instrument kindly provided by Dr. N. Furlong. It comprises a climatized chamber that allows provision of a saturated atmosphere. The recording of the water droplet shape is done by a video camera and the shape digitized. Using this apparatus, drops were observed over a period of five minutes; if by then no change had occurred, it was inferred that the plasma films did not absorb water. Such films also did not change their interference colouring on the metallized substrate following application of a drop of water and careful removal by a tissue after a few minutes, indicating no noticeable swelling.

Scanning tunneling microscopy (STM) was performed on a custom-built STM unit provided by Dr. B. A. Sexton (Division of Materials Science, CSIRO). Polymer samples were prepared for STM study by sputter deposition of a thin conducting layer of gold or platinum. ATR-IR spectra were recorded on a Mattson Alpha Centauri spectrometer with the samples pressed against a KRS-5 crystal. Grazing angle (5° incident angle) FTIR spectra were obtained in the noncontact reflecting mode using a Spectra-Tech FT-80 accessory. X-ray photoelectron spectroscopy (XPS) analysis was performed with a VG Escalab V unit using Al K_a radiation at a power of 150 W. Elements present were identified by survey spectra. Elemental surface compositions were obtained from numerically integrated peak areas and applying standard sensitivity factors. Individual peaks were then analyzed at high resolution to determine the contributions due to different bonds. Standard techniques were used for Gaussian curve fitting.

Human endothelial cells derived from umbilical arteries (HUE cells) were established by the methods of Jaffe et al. [15]. For other experiments, fibroblast cells from human dermal tissue (HDF cells) were used. The HUE cells were confirmed to show positive immunostaining for von Willebrand's factor (factor VIII-related antigen). The HUE cells were maintained in a humidified atmosphere of 5% CO_2 in air at 37°C in a medium that consisted of Medium 199 (Flow Laboratories) supplemented with 20% (v/v) foetal calf serum (Cytosystems, Sydney), 60 μg/ml endothelial cell growth supplement (Collaborative Research), 100 μg/ml porcine heparin (Sigma Chemical Co.), 60 μg/ml penicillin (Glaxo), and 100 μg/ml streptomycin (Glaxo). The cells were routinely passaged using trypsin-EDTA and grown in tissue culture polystyrene (TCP) flasks (Corning), which had been incubated with a solution of 40 μg/ml bovine plasma fibronectin (Sigma Chemical Co.) in phosphate-buffered saline at 37°C for at least 1 hr prior to inoculation with cells. The cells were used in these experiments at passage numbers five to seven. Cell attachment was determined by examining the number of cells that spread onto the surface, showing deviation from a round, phase bright morphology and producing cytoplasmic extensions onto the substratum following one and successive days of culture. Cell cultures were fixed with 2.5% glutaraldehyde in phosphate buffered saline, then stained with Eosin Y (Gurr). Randomly selected fields were photographed under epifluorescence UV.

3.4 PLASMA SURFACE TREATMENTS

Plasma treatment of PTFE and FEP in the process gases (air, oxygen, water vapor, and argon) in all cases proved very effective for surface modification of these normally highly unreactive polymers. The effects of surface modifications cannot normally be seen by the naked eye on these samples, except when relatively high power levels and long treatment times are used, in which case yellowish or brownish discolouration of the samples and often some deformation occur. However, even for samples that appear visually unchanged, dramatic changes can be seen when wetting the samples with water or other polar liquids: while unmodified PTFE and FEP are very hydrophobic and thus the liquids bead markedly, plasma-modified samples were easily wettable, with good and uniform spreading of the liquid drop. This wetting behaviour can be quantified by the rapid and convenient method of determination of the contact angle at the liquid/air interface of a drop on the polymer surface. Advancing, sessile, and receding contact angles provide complementary information on the ease of wetting and a measure of the polar and nonpolar surface areas. Contact angles are the first means of assessing plasma treatments of PTFE and FEP samples; even when samples are visually unchanged, marked decreases in contact angles are invariably observed. Figure 3.1 shows typical data documenting the increase in the wettability of the polymer surface by distilled water following plasma treatment of FEP in a plasma ignited in a water vapour atmosphere at various lengths of exposure of the samples to the gas plasma.

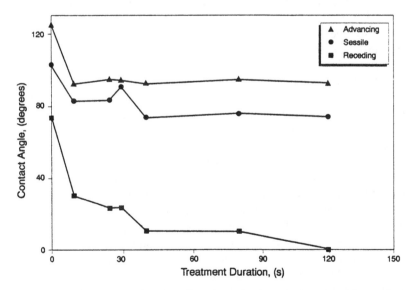

Figure 3.1 The sessile (●), advancing (○) and receding (×) contact angles of FEP samples: unmodified, clean FEP ("zero treatment time") and exposed to a water vapour plasma for various lengths of time.

The decrease in contact angles consequent on plasma treatment depends also on the plasma power level applied; the above data were obtained at relatively low power, but it is evident that sufficient wettability has been conferred. The advancing angle provides an indication of the hydrophobic surface domains which oppose the advance of the liquid front over the surface, while the receding angle is indicative of polar surface domains, which prefer to retain contact with the polar liquid rather than with the nonpolar air. The difference can be used to estimate the relative densities of these segments [18], but a number of rather large uncertainties are involved. For the present review, the data are used merely to document that polar groups have been attached to the polymer surface. Detailed analysis of contact angles measured using a number of liquids with various polarities allows calculation of the dispersive and polar contributions to the surface energy; for these treatments on FEP and PTFE, it can thus be shown that it is the latter component that is increased by plasma treatment [16], which is consistent with the expected attachment into the surface layer of polar groups.

Other process gases produced substantially the same dependence of the decreases in contact angles as a function of treatment time. It is easier from a process point of view to use air or oxygen administered from the ambient atmosphere or a cylinder, respectively, via a flow meter to the chamber, than to use water vapour, which needs to be obtained from the vapour phase on top of the liquid placed in a suitable vessel. The rate of boiloff of water vapour from the liquid restricts the rate of flow and thus processing. However, it appears that water vapour plasmas are more gentle on the polymer. All these plasma atmospheres attach oxygen containing polar groups to the polymer surface, and these groups confer wettability, but the chemical composition of the treated surfaces is not the same for all process gases. Water vapour appears to attach more hydroxyl groups (detailed surface analysis is in progress) whereas air and oxygen attach more species with higher oxidation states. Oxygen may attach some groups, which then promote continuing chain oxidation, leading to some oxidative degradation of the polymer surface layers. In other words, the plasma needs to be administered such that it avoids or reduces reactions with possible long-term detrimental effects akin to those of conventional polymer degradation reactions, which can reduce the service life of the polymeric article. Recent data have shown that over weeks following plasma treatment, further oxygen uptake occurs [18], and the nature and effects of these reactions are currently under study.

It is important to emphasize the short duration of plasma treatments needed for effective surface modification of these polymers, which are inert to all but the most severe of conventional chemical reactions. Even at the low power levels used (12 to 30 W), treatment times of 10 sec were sufficient to confer wettability. These data were obtained with stationary samples placed on electrodes, and thus the time it takes from switch-on to the establishment of a stable plasma atmosphere defined the lower limit for treatment times. Further experiments are required in order to define the minimum residence time needed for effective surface treatment in a stable, equilibrated plasma; this time may be less than a few seconds. Plasma treatment times of many minutes are sometimes used in industrial plasma applications; it is considered that such extended treatment times are unnecessary and probably even detrimental. Overtreated samples often turn yellowish and brittle; it appears from preliminary data that such overtreatment actually has similar effects to those seen in the oxidative degradation of polymers. Therefore, it is necessary to be careful in optimizing surface properties without sacrificing service life of polymeric articles. Experimentation with various internal fittings and electrode geometries produces plasma discharges of various densities and shapes, and it appears that the duration needed for effective treatment depends strongly on the geometry of the plasma and the experimental parameters. For industrial applications, treatment times can probably be reduced considerably from present practice by proper equipment design.

The next question is, what exactly the chemical reactions are that the plasma gases produce on the polymer surface. Even simple plasma atmospheres such as that from oxygen gas contain a number of reactive species, which are atoms, ions, radicals, new and usually unstable molecules produced in the plasma discharge, electronically excited species, and free electrons. The relative amounts depend on the plasma parameters and are not easy to quantify. Even harder is the prediction of the various reactions between these species and a polymer; thus, the chemical effects of plasma surface treatments are currently not predictable, and also, they are not chemically very selective. Previous studies have found that plasma exposure typically produces several different new groups on polymer surfaces [4,10,11,17,18]. Thus, detailed surface analysis is required, but this is a considerable challenge not only for plasma-treated surfaces, but for polymers, in general [19], and is the subject of much ongoing work worldwide.

Likewise, study of the chemical composition of our plasma-treated fluorocarbon polymer surfaces proved less than straightforward. For polymers, infrared spectroscopy usually is the chemically most informative "surface" spectroscopy method, since it indicates chemical bonds present. However, transmission IR spectra were identical for treated and untreated FEP samples; PTFE samples could not be used as they are opaque. ATR-FTIR spectroscopy also failed to show spectral differences. The ATR penetration depth is of the order of 1 to 2 μm at incident angles of 30° and 45°, whereas previous work using XPS analysis of plasma-treated polyethylene found that, for that polymer, the effects of plasma modification were limited to the production of new chemical groups over a depth of a few nm only [10]. Assuming that plasma treatment is similarly shallow on fluorocarbon polymers, ATR-FTIR lacks the sensitivity to detect these exceedingly small amounts of new groups.

X-ray photoelectron spectroscopy (XPS or ESCA) possesses an analysis depth of about 3 nm for polymer samples; for instance, in a study by Gerenser [10], this technique provided valuable information on the surface chemical gorups produced by plasma treatments similar to ours but on polyethylene. XPS analyses confirmed that the changes in wettability of PTFE and FEP also were due to chemical changes on the polymer surfaces rather than to surface roughness changes (which can also alter contact angles), as shown by a new peak assignable to oxygen [16]. The increased wettability following plasma treatment is thus assigned to the substitution of some of the $C-F$ bonds by polar oxygen-containing chemical groups. However, the oxygen uptake was less than the fluorine loss, indicating that some of the carbon radicals produced by homolytic scission of $C-F$ bonds reacted with each other to form interchain crosslinks or double bonds. Under a given set of plasma conditions, the O/C ratio was higher for plasma-treated FEP samples than for plasma-treated PTFE samples, suggesting that plasma reactions with surfaces are quite sensitive to the relatively small differences between these two polymers. The chemical identity of the oxygen-containing groups produced could not be clarified from XPS spectra alone; the C 1s spectrum was very broad and asymmetric, indicating that several new groups existed on the modified surfaces, and their contributions overlapped to an extent that prevented meaningful curve fitting, in particular on account of the various secondary shifts that need to be considered. Detailed analysis of plasma-treated fluorocarbon polymer surfaces requires the development of additional methods such as XPS combined with specific chemical tagging reactions; this is the subject of ongoing work here and at a number of other laboratories.

For polyolefins, it has been found that plasma surface oxidation mainly breaks $C-H$ bonds and introduces hydroxyl, ether, peroxide, carbonyl, and carboxyl groups onto the surface [17]. Plasma treatment produces a considerable number of carbon-centered radicals on the polymer surface; these radicals can react with each other to form crosslinks and $C=C$ double bonds, or they can react with a species that arrives from the gas phase. If no suitable species for reaction is available from the gas phase, as is the case when argon gas is used, then many radicals are still available when the reactor is vented with air. These remaining radicals will react with oxygen and water vapour to attach additional groups. Atmospheric nitrogen is far less reactive than oxygen [10]; hence, little or no N is observed by XPS analysis of plasma-treated surfaces.

There is also need to emphasize the requirement for clean surfaces for plasma treatment; when impurities are present the plasma gases may chemically modify these impurities rather than the polymer surface. Subsequently, the impurities may become partly or wholly lost, and the treated polymer surface is no longer wettable. Oxidative plasmas are used in plasma surface cleaning for the removal of impurities, in particular in the semiconductor industry, but the aim there is different in that no permanent modification of the substrate itself is needed. At short treatment times, however, impurity removal may be incomplete. Ultrasonication in ethanol wash samples and XPS verifies cleanliness prior to plasma treatment [18].

The effects of plasma treatments of fluorocarbon polymers are also readily observed in experiments involving the attachment and growth of cells onto polymer surfaces immersed in cell culture media. Cells evidently are incapable of attaching to pure fluorocarbon surfaces, and on the isolated sites where they do attach (these sites may be small defect sites or remaining impurities at a very low level), cells then show great reluctance to grow sideways but instead, stack on top of each other. After plasma surface oxidation, however, the attachment and growth is much more effective [13], although still not as good as on the commercially available reference material—tissue culture polystyrene (TCP). TCP is thought to be fabricated by oxidative plasma treatment as well. Thus, oxidative plasma treatment could, in principle, be adequate for surface modification of biomedical devices, which, in order to possess good mechanical properties, needs to be fabricated from fluorocarbon bulk material.

However, plasma-modified FEP samples possess surfaces that slowly change with time; the wettability conferred by plasma surface treatment decreased, and changes were also observed in the XPS analysis data [18]. Figure 3.2 reproduces the increases of the sessile contact angles, with water as a function of storage time, of two argon plasma-modified FEP samples.

Similar decreases in the wettability with time have previously been observed for plasma-treated hydrocarbon surfaces [1]. The phenomenon is not restricted to plasma-modified surfaces; polymer surfaces treated by corona discharge show much faster decays of the treatment effects. The surface of polymers generally displays a certain mobility because thermal motions of chain segments cause reorientation of the near surface layers. This mobility is responsible for the decay of the wettability; chemical groups introduced by surface modification may not reside permanently at the surface, but slowly become buried inside the polymer. The driving force is the reduction of surface energy when an increasing percentage of nonpolar groups is in contact with air. However, various degrees of reversal with varying treatment have not been reported before.

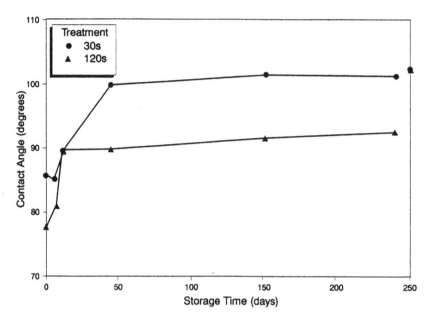

Figure 3.2 The sessile contact angle between a drop of distilled water and argon plasma–modified FEP samples, as a function of time of storage: (●) treatment duration 30 s; (×) treatment duration 120 s. The contact angle with unmodified FEP is plotted on the right-hand axis to demonstrate the extent of surface reversal.

It is evident that, for sample treated for 30 s, there is a virtually complete loss of the treatment effects within fifty days, whereas the sample treated for 120 s loses the treatment effects to about 60% within the same time and then shows a much slower rate of loss. The data of Figure 3.2 may be interpreted by assuming that the plasma treatment produces basically *two* populations of oxygen-containing groups. One population comprises groups that are attached to mobile chain segments and can diffuse into the polymer relatively rapidly, while the other population contains oxygenated groups that are attached to carbon atoms in the vicinity of crosslinked structures in the surface layer. Groups attached at areas that have a locally high crosslink density will reorient much more slowly and are responsible for the semi-permanent component of the wettability. XPS data showed that the oxygen content of the two samples did not differ much, in accordance with the conclusion from plots such as that in Figure 3.1, which shows that the substitution of fluorine by oxygen-containing groups is essentially complete within a short time of plasma treatment. However, the fluorine content of the sample treated for 120 s is smaller; this is in agreement with a higher degree of crosslinking, which would decrease the mobility of a larger fraction of the surface layer and thereby prevent a larger amount of the oxygen-containing groups from taking part in the more rapid component of the surface reorientation. The sample treated for 30 s apparently has very few, if any, surface areas with sufficiently high crosslink density to immobilize their chain segments for a number of weeks. Longer plasma treatment times thus do not substantially increase the extent of surface oxidation but continue to alter the polymer surface by producing crosslinking. This situation appears unique to the fluorocarbon polymers because hydrocarbon polymers appear to continue to undergo oxygen incorporation with longer exposure.

The above example serves to demonstrate the difficulties in unravelling in detail the chemical effects of plasma exposure on polymer surfaces. However, even the above analysis is somewhat tentative because, in spite of the relatively simple polymer and plasma gas phase used, close inspection reveals unexpected additional complexity: XPS analysis showed that the chemical changes within the XPS analysis depth (~ 3 nm) were not only due to the plasma exposure, but even more complicated because reactions continued to occur during storage, and their effects were superimposed on the effects assigned to mobility [18].

This mobility of the surface raises concerns. Firstly, it would be better to use stable, well-characterized surfaces for the study of interactions with proteins and cells; surfaces that reorient can present unpredictable and ill-defined chemical compositions to the environment, which itself modulates these rearrangements. Secondly, from a practical point of view, when considering biomedical applications of polymers, the question arises whether such mobile surfaces will be able to maintain the ability to exert the desired degree of control over interfacial interactions given that the surface reorientation may be reversible. On storage in air, which is hydrophobic, a device comprising a surface-treated fluoropolymer can be expected to change towards a hydrophobic, fluorocarbon surface by motions of the mobile chain segments. Following implantation the biomedical device suddenly "sees" the hydrophilic environment of the host body, which causes again a driving force working towards minimization of the surface energy, but this time by re-emergence of the buried polar groups. Accordingly, chain motions will start to gradually produce a more hydrophilic surface. The biomedical effects associated with such rearrangements of polymer surface layers while implanted are not known or predictable at this time. However, given that protein adsorption is very sensitive to surface composition and polarity, substantial effects might be expected.

3.5 PLASMA POLYMERIZATION

The partial loss described above of the treatment effects is undesirable both for fundamental research and for potential applications in biological interactions, but chain mobility is a fundamental property of polymers unless they are extensively crosslinked. Plasma surface treatments, which do not deposit additional coatings but only attach new functional groups, also generate some surface crosslinking and thus reduce the rate of chain reorientation. This may not be sufficient for freezing the surface composition, as for instance in the above example. A higher degree of crosslinking can be achieved at higher plasma power and/or longer exposure, but such measures may induce other adverse effects (overtreatment, excessive heat load). When extended stability of the surface composition is required, a procedure is desirable that also inherently restricts mobility at the same time as it produces a surface with different chemical groups. This can be achieved by making use of depositing plasma atmospheres. Instead of altering the chemical groups on the fluoropolymer surface, a thin plasma polymer coating is deposited onto the environment. The surface properties of the composite are then determined solely by the chemical constitution of the surface of the plasma polymer. The growth of cells, for instance, will then be governed only by the plasma polymer coating, without regard for the bulk polymer underneath. A further advantage is that, once a plasma polymer coating has been optimized, it can be applied to any commercial bulk polymer to adapt it to a particular application, whereas the chemical effects of (nondepositing) plasma surface treatments are determined by both the plasma gas and the substrate polymer and therefore need to be optimized for each new bulk polymer. Plasma polymerization can also be used to deposit thin polymeric coatings onto metallic and ceramic substrates in order to provide application specific surfaces. While plasma polymerization is thus very attractive for its ease of transfer to various substrates, it is inherently slower and more complex to optimize and control than plasma surface treatment.

Plasma polymer coatings characteristically have a random structure built up from a variety of chemical entities derived from the organic vapour. Typically, random activation causes few linear chain segments analogous to conventional polymerization reactions to be incorporated, and the result is a substantial degree of crosslinking, which varies with monomer and plasma conditions. As a result of the crosslinking, plasma polymers generally possess little or no flexibility of polymeric segments. Accordingly, contact angles on plasma polymer surfaces have been found to be invariant with time or vary slowly [1]. Thus, the production of a crosslinked thin "skin" layer, by plasma polymerization, onto PTFE and FEP appeared promising for fabricating samples with stable surfaces, which also do not reorient as the environment changes.

However, while stability may be achievable, the fabrication of *predictable* surface compositions remains a challenge. Even well-established and chemically apparently simple surface modification methods such as acid etching produce surface compositions that usually are neither predictable nor easy to analyze [19]. Plasma surface treatment in simple gases still is insufficiently understood, and in the more complex plasma phase established in the vapour of an organic compound, the same unpredictability applies to an even larger degree. Plasma polymerization proceeds by various, nonthermal reactions that a monomer molecule can undergo following activation in the plasma. Plasma polymers usually contain significant diversity of the bonding structure and do not contain the regular, periodic simple structural elements of conventional polymers. The surface of plasma polymers can be expected to contain several different chemical groups, incorporated by the various reactions; this structural diversity is analogous to the chemical diversity obtained by plasma and corona surface treatment although derived from different reaction pathways. As for surface-modified polymer samples, prediction is unreliable at present, and it is necessary to analyze plasma polymers in detail so that correlations between fabrication parameters and the composition can be established. Again, such analysis is complicated by the presence of several chemical groups containing similar combinations of elements. In particular, several oxygen-containing groups are typically observed.

Currently available, established methods for the analysis of polymer (see, e.g., an excellent review by Ratner et al. [20]) are not capable of fully determining the nature and density of the various chemical groups on surfaces, be it of plasma polymers or polymeric materials in general. However, the recent application to polymers of static secondary ion mass spectrometry (SIMS) [21], which has very high surface selectivity and can provide structural information to a much larger extent than the limited bonding information available from XPS, has provided valuable data on a few systems and will, with increasing availability and in conjunction with XPS and IR methods, allow more comprehensive analysis. Another important area of research is the tagging of target groups with specific labels which, hopefully, attach selectively. Several groups are active in this area, but it has become increasingly clear that the avoidance of side reactions is more difficult than initially anticipated [22,23]. In the absence of comprehensive surface characterization, full correlations cannot be established between plasma parameters and plasma polymer composition but are limited to semi-quantitative work.

This work has partly been based on the premise that, while plasma polymerization cannot be fully chemically selective for fundamental reasons, under suitable conditions it might be possible to reduce the often extensive monomer breakage and randomization of bonds, so that it might be able to incorporate into the plasma polymer a considerable fraction of a desirable functional group from the monomer. Probably relatively mild conditions should favour the retention of desired groups by reducing fragmentation of the monomer, which can be extensive under high plasma power. It is then necessary to confirm that an appropriate density of such functional groups is also located at the surface—it cannot be taken for granted that the surface composition of a plasma polymer is the same as the "bulk" (say, 50-nm coated thickness) composition. Furthermore, the degree of surface mobility needs to be investigated. Regarding chemical selectivity in plasma polymerization, it might be possible to select monomers that show stronger preferences for proceeding along a smaller number of reaction pathways, but in the absence of plasma analysis of organic vapour plasmas, intuition will be needed to guide such work.

A series of ultrathin (mostly ~50-nm thick, some up to 150 nm) plasma polymer coatings has been produced from a range of monomers. Alcohol and amine monomers have been used with the intention of fabricating plasma polymer surfaces rich in amine and hydroxyl groups because these groups are ubiquitous in biological environments and thus such polymer surfaces might demonstrate good biocompatibility. Other surfaces were selected for other reasons. The plasma polymerization of organosiloxane monomers, in particular hexamethyldisiloxane (HMDSO) can be used to fabricate a crosslinked, plasma polymer analogue of the conventional polymer polydimethylsiloxane, which is well tolerated in biomedical environments. Other plasma polymers have been made so that a wide range of hydrophobic/hydrophilic surfaces and a variety of surface groups were available for biomedical testing. Such testing might then show up the factors of importance in fabricating biomedically optimal polymer surfaces.

Prior to biomedical testing, however, it was important to ascertain that the plasma polymerization process produced highly conformal coatings. It is well known that cells react sensitively to the topography of a surface; in order to study the biomedical effects of various chemical groups on polymer surfaces, the surface modification procedure needs to produce minimal changes to the topography. This point has often been overlooked, and procedures such as radiation grafting invariably produce structured coatings that do not readily allow separation of chemical and topographical effects. The present thin coatings required a technique that can assess surface topography on the nanometre scale. Scanning tunnelling microscopy (STM) is ideally suited for this purpose but requires a conducting sample surface. This was achieved by sputter coating a thin Pt layer onto the samples. Reference FEP samples were found to have a surface roughness typically on a scale of a few tens of nanometres. After coating with a plasma polymer, identical values within experimental error for the mean surface roughness were found, and visual inspection of STM plots showed no differences. An example of an STM surface contour plot is reproduced in Figure 3.3; the plasma polymer coating in this case was deposited from dimethylformamide (DMF) monomer vapour.

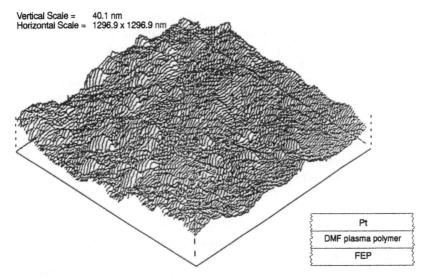

Figure 3.3 STM surface contour plot of the multilayer system.

The metallized Kapton substrates were much smoother than FEP and allowed more detailed assessment of possible surface topography effects due to plasma coating. Because of the presence of surface metal oxides, the reference samples also needed to be coated with a Pt thin film to obtain better conduction. Figure 3.4 reproduces an STM surface contour plot of

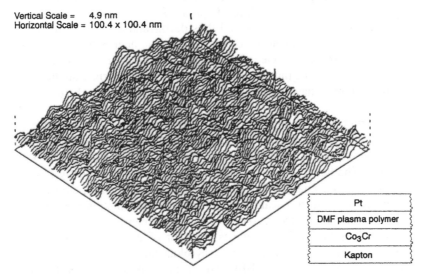

Figure 3.4 STM surface contour plot of the multilayer system.

a plasma polymer coating on metallized (Co_3Cr) Kapton tape. Comparison with the plot obtained on the $Pt/Co_3Cr/Cr/Kapton$ sample (not shown) indicates that the plasma polymer did not significantly increase the mean surface roughness on a nanometre scale.

The factors limiting the resolution in STM of these polymer samples were also investigated. Radiofrequency sputter coating of Pt onto molecularly cleaved mica gave a value of 3.0 nm for the z-scale on a 100-nm square scan. Sputter coating of Pt in a dc discharge onto mica provided a coating with only very slightly higher mean surface roughness but the distribution of grain sizes was wider. The z-scale on a 100-nm square scan was 3.6 nm. Radiofrequency sputter coating of gold onto mica showed a somewhat larger grain structure again; the z-scale on a 100-nm square scan was 4.9 nm.

Comparison with the $Pt/Co_3Cr/Kapton$ sample indicates that the microcrystalline structure of the radiofrequency sputtered Pt film is not alone responsible for the surface roughness; the observed roughness is partly due to the Kapton tape and partly to the Pt layer. These effects define the lower limit to observations on plasma polymer coatings. Comparison of STM data from the $Pt/Co_3Cr/Kapton$ reference sample with the same system plus a plasma polymer, e.g., that of Figure 3.4, demonstrates that the effects, if any, due to the plasma polymer cannot be quantified with sufficient accuracy for most plasma polymers studied because the data (z-scale and rms surface roughness) are identical within experimental variability. The experimental variability appears not to be due to variability of the plasma coating but due to slight local variations in substrate topography; different areas on $Pt/Co_3Cr/Kapton$ reference samples give slightly different values for the z-scale and thus deconvolution is not sufficiently accurate. The STM contour plots recorded on a number of plasma coatings thus do not provide absolute information on the plasma polymer itself, but they give the valuable (and not previously reported) information that these plasma polymers have an excellent thickness uniformity to better than a few nm.

The structure of the conducting (Pt or Au) coating is only of relevance when attempting to achieve a very high resolution, and differences between these coatings could only be observed on cleaved mica and the Kapton tape. The latter is the smoothest polymer material presently studied. For other commercial polymers, such as the FEP tape used, their surface roughness is of such a scale, typically tens of nanometres rms, that the grain size of the conducting coating is no longer relevant. However, the absolute values for surface roughness are not of primary interest in this study; a main aim is to demonstrate, by plots such as Figure 3.3, that the materials (mainly, reference FEP and plasma polymers on FEP) used in testing involving cell attachment and growth do not significantly differ in surface topography, so that the observed differences in the behaviour of cells can unambiguously be assigned to *chemical* differences.

Most plasma polymer coatings of ~50- to 150-nm thickness were, in this way, shown to be highly uniform and conformal to the substrate. However, exceptions existed, and Figure 3.5 shows the STM surface contour plot of an allylamine plasma polymer to have a surface roughness markedly in excess of that associated with the FEP substrate itself. This is often the case when plasma polymerizing allylic and acrylic monomers; the double bond allows the conventional radical polymerization to take place in addition to the plasma process. Thus, fast reactions occur in the gas phase, and oligomers and polymer particles are obtained, which diffuse to the surface and become incorporated into the growing plasma coating. Freeze-fracture cross sections investigated by scanning electron microscopy showed spherulitic inclusions and protruding hemispheres in such coatings.

The composition of the plasma polymers was studied by spectroscopic analysis using IR and XPS techniques. Transmission IR of plasma films on FEP was not sensitive enough because the much thicker substrate obscured the bands excessively due to the plasma coatings; subtraction did not provide reliable spectra unless coating thickness values much larger than the ones we typcially prefer to work with were used. Thicker plasma coatings often accumulate substantial internal stress. ATR-IR provided limited information and poorly reproducible spectra with the typical high

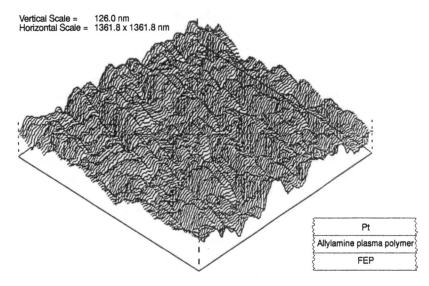

Vertical Scale = 126.0 nm
Horizontal Scale = 1361.8 x 1361.8 nm

Pt
Allylamine plasma polymer
FEP

Figure 3.5 STM surface contour plot of the multilayer system.

noise in the high-frequency region indicative of inadequate surface contact between the rather inflexible plasma polymer and the crystal. Furthermore, the signal intensity was exceedingly low due to the thinness of the plasma polymer films. Much better spectra were obtained with either the grazing angle accessory or in the specular reflectance mode, both of which also offered the advantage of noncontact measurements. However, these methods demand a highly reflective, metallized substrate underneath the thin plasma coating. Thus, plasma polymers on the Co_3Cr/Kapton substrate were used for such analysis. Since the contact angles for plasma films deposited onto FEP and metallized Kapton were the same, it was assumed that the film composition is identical and independent of the substrate. For grazing angle IR studies, the low incident angle (5°) also requires a high degree of smoothness of the metallized substrate surface in order to produce satisfactory spectra; the custom metallized Kapton base was well suited to this purpose.

Infrared spectra provide an indication of the chemical groups present in the plasma polymer structure, but quantitation is an unsolved problem in specular reflectance and grazing angle FTIR. The relative band intensities vary with incident angle, polarization, etc., and these effects are not understood at present. Neither are quantitative spectra from reference polymers available. Thus IR methods were used only for qualitative analysis of plasma polymers, although in some cases it is believed that comparison of the observed intensity of two bands can be used to provide a tentative conclusion as to which group is more prevalent.

In principle, XPS analysis can be used for quantitative work, but for polymers the problem arises that straight XPS cannot differentiate between a number of chemical group of interest. The ability to identify particular groups, which is so attractive in IR, exists only to a limited extent in XPS. Most polymers, even apparently simple conventional ones, possess considerable diversity of the chemical structures present on the very surface; the composition of the surface usually differs considerably from that of the bulk for various reasons [20]. Identification of the surface groups has been the subject of much work, and for separation of similar groups, additional labelling reactions have been utilized, but there are still concerns regarding their selectivity [22], and work is continuing on the development of new and improved methods of surface analysis [22,23]. The quantitative analysis of plasma polymers by XPS is similarly difficult.

A number of plasma coatings have been analyzed, some of which were made for nonbiomedical research purposes, to the extent possible with the equipment and methods currently available and have in this way gained a reasonable semi-quantitative understanding of the composition of some of the plasma polymer films. Examples are studies by combined IR and XPS analysis of plasma polymers from hexamethyldisiloxane [24] and of plasma polymers from alcohol, amine, and amide monomers [25]. Literature data are of limited value in such an analysis because the results of plasma polymerization are known to depend markedly on the apparatus geometry. Results from other laboratories may give qualitative indications of preferred pathways of plasma polymerizations. However, even the relative importance of reactions may shift markedly with the design of the plasma glow zone; therefore quantitative analysis needs to be done on samples fabricated on our equipment. Here a brief analysis is given, which is presented in more detail in Reference [25], of a plasma polymer made from heptylamine monomer to document typical results and capabilities.

The IR spectrum contained an absorption band assigned to an NH vibration as a broad band between 3100 and 3600 cm^{-1}. C$-$H and amide bands were also evident. These structures are consistent with a polymer primarily based on a hydrocarbon backbone, which is derived from the *n*-heptyl chain, with functional groups derived from the amine substituent. The amide groups presumably were produced by reaction between atmospheric oxygen and radicals adjacent to amines, radicals that were not dissipated in the film formation process and remained available for reaction on venting the plasma reactor with air. The relative intensity of the NH band and the amide band (1691 cm^{-1}) was clearly not consistent with a pure amide; the former was too intense, suggesting that some of the NH was in the form of amine. Compositional analysis by XPS gave an N/C ratio of 0.08 and an O/C ratio of 0.04. This is consistent with a mixture of amine and amide, with at least 50% of the N contained in the former. Some of the O may not be in the form of amide but as carbon-oxygen groups (hydroxyl, carbonyl, etc.). As to the amine groups, it is not yet possible to determine the relative amounts of primary, secondary, tertiary, and quaternary amine groups, but work is in progress.

The N/C ratio of the plasma polymer can be compared with the N/C ratio of 0.14 of the monomer; it shows that a substantial percentage of the amine groups became detached from the hydrocarbon backbone during the plasma polymerization of the heptylamine molecules. This is in qualitative agreement with earlier work that suggested a high susceptibility of the C$-$N bond in amines towards breakage in a plasma. However, whereas previous studies have suggested that the functional groups are lost before much polymerization occurs, some success has been achieved in partly combating this facile abstraction and retained more than half the amine groups (some of them apparently converted to amides). It is concluded that it is possible by judicious choice of plasma parameters to steer the plasma reactions partly away from abstraction of heteroatom groups and to activate monomers mainly by C$-$H breakage, thus fabricating hydrophilic plasma polymers by the formation of new C−C bonds, as in the plasma polymerization of alkanes and the simultaneous incorporation of pendant amine groups.

Substantial retention of the functional group under suitable conditions can also be obtained in the plasma polymerization of saturated alcohols; for instance, the grazing angle IR spectrum of an ethanol plasma polymer film showed a marked hydroxyl band as well as a carbonyl band [25]. The latter derive from the unavoidable oxidation reactions. The relative band intensities suggest a higher density of hydroxyl than carbonyl groups. XPS analysis is in agreement with the IR indication of a high hydroxyl density. The O/C ratio was 0.30; comparison with the value of 0.5 for the monomer again shows that, although some of the functional groups were abstracted during polymerization, cleavage of the C−O bond is not a major pathway under our plasma polymerization conditions. Again, using relatively mild plasma conditions, it has been possible to prepare a film with a high concentration of a desired chemical group.

Yu and Marchant obtained hydroxyl rich surfaces by a two-step process involving plasma polymerization of N-vinyl-2-pyrrolidone and reduction of the predominantly produced carbonyl groups by sodium borohydride [26]. The present approach, however, is the use of appropriate monomers and plasma conditions to take the simpler, one-step route to hydroxyl-rich surfaces. Analogously, previous work found that the plasma polymerization of amines produced coatings with a large part of the N incorporated in the form of amides [27]; again, present results show that the relative amounts of amines and amides can be tipped in favour of the former under suitable conditions. The previous study used the unsaturated compound allylamine as monomer, whereas we have used saturated amines; perhaps the double bond of allylamine is more susceptible to subsequent oxidation.

To date, there have been few studies of plasma polymerization aimed at selective production of specific chemical groups. Both the alcohol and amine plasma polymerization results suggest that there is considerable scope in optimizing monomers and plasma parameters for compositional control of plasma coatings.

3.6 SURFACE ANALYSIS OF PLASMA POLYMERS

Currently several applications of plasma polymers are in progress−wetting, adhesion, biocompatibility, corrosion protection−but an important common factor in much of the work is the composition of the topmost molecular layers. Molecules approaching from the environment experience only the chemical groups in close proximity to the interface because the range of the interfacial forces is exceedingly small. Such interfacial interactions govern adhesion, wetting, and biocompatibility because, for instance, biomolecules such as proteins from the host body will first and foremost interact with the topmost layer of the polymer on approach, long before in-diffusion takes place. The range of the interfacial forces depends somewhat on the type of force (hydrogen bonding, ionic, hydrophobic, etc.); several are relevant; and accurate values cannot be given, but there always is a steep dependence on distance.

It appeared reasonable to presume, based on the current, limited understanding of the plasma gas processes in organic vapours, that the composition of the surface of plasma polymer films may be identical to that of the "bulk," but definitive evidence is not available. In fact, preliminary data on one type of plasma polymer suggest the contrary [28], but it is not known whether this is a special case. Therefore, methods are being developed to analyze for the presence and density of various groups within the top surface layer of plasma polymers.

The analysis depth of the "surface" technique XPS is about 3 nm for polymers, a value that is still substantially larger than the range of interfacial interactions, and thus XPS is still essentially a "bulk analytical" technique for our purposes. XPS spectra taken at lower emission angles probe closer to the surface and provide a depth profile for the atoms or groups of interest. However, again, this needs to be complemented by labelling methods that selectively identify one group. A number of the labelling reactions described in the literature rely on vapour molecules as reactive labels; these gaseous compounds appear capable of penetrating into polymers and thus perform their labelling into the depth of the polymer, with combined diffusion/reaction limited profiles [22]. This makes analysis and quantification of the very surface extremely difficult. It is necessary to develop methods using labelling, reactive molecules, which do not penetrate into the polymer and thus selectively label only the target groups in the topmost polymer layer. The easiest way of avoiding diffusion into the polymer appears to be the use of relatively bulky molecules that are insoluble in the polymer, and efforts have made use of dyes with such properties [23,25]. However, the exclusion of side reactions and identification of nonspecific adsorption masking true covalent labelling are considerable challenges, and it has not yet been possible to give a complete analysis of a plasma polymer surface. Work in this area will also be useful for other polymers and other surface modification studies. At this stage, the labelling aspect of surface analysis is in need of much further work.

Work in these laboratories so far has concentrated on developing a method for the quantitative analysis of primary and secondary amine groups on surfaces of plasma polymers made from amine monomers. Initially, good results were obtained using a fluorescein dye [25], but side reactions appeared to occur on some sample surfaces, leading us to other probe molecules. Preliminary results from this subsequent work were presented in another contribution at the conference [23] on surface engineering, from which this book is drawn.

3.7 BIOMEDICAL PERFORMANCE OF PLASMA POLYMERS

The ability of polymer surfaces to support the attachment and growth of fibroblasts from a cell strain prepared from human dermal tissue (HDF cells) and of human endothelial cells was determined. The plasma polymer films used were deposited from the monomers hexamethyldisiloxane (HMDSO), hexamethyldisilazane (HMDSA), toluene, n-hexane, methyl methacrylate (MMA), ethyl methacrylate (EMA), methanol, ethanol, and isobutanol. The attachment of HDF cells to the plasma-treated surfaces were compared with that to unmodified fluorinated ethylene-propylene (FEP) and to "tissue culture" polystyrene (TCP).

HDF cells showed only very poor attachment to FEP, and the attached cells had a spindle-like morphology that is characteristic of poor cell attachment. The HDF cells on FEP showed very limited cell growth and lifted off the surface after two to three days' culture. On the TCP surface, the HDF cells showed both good cell attachment and growth, achieving a cell density of greater than 50,000 cells per square cm during the three-day culture, following seeding with 12,500 cells per square cm. The surfaces treated with toluene, n-hexane, MMA, EMA, methanol, ethanol, or EMA plasmas supported HDF cell attachment and growth to an extent that was clearly improved over the untreated controls. The rate of cell growth on the surfaces was calculated for each surface on each of days one, two, and three of culture and is presented in Figure 3.6 as a plot of cell growth rate as a function of cell density.

Most of the plasma surfaces supported HDF growth at a level that was approximately equivalent to that of TCP. It can be seen that for these surfaces (open squares; compared with TCP, diagonal square), the rate of cell growth decreases as a function of cell density and the rate of cell growth diminishes markedly above a density of 40,000 cells per square cm. The relatively small scatter of the data (at any particular cell density value) indicates that the toluene, n-hexane, MMA, EMA, methanol and ethanol plasma surfaces did not differ significantly from one another or from TCP in their ability to support HDF cell growth. In contrast, each of the FEP modified with HMDSO and HMDSA (closed circles) were relatively poor at supporting the attachment and growth of HDF cells, and the data for these surfaces and for unmodified FEP (closed triangles) fall as a separate group in Figure 3.6.

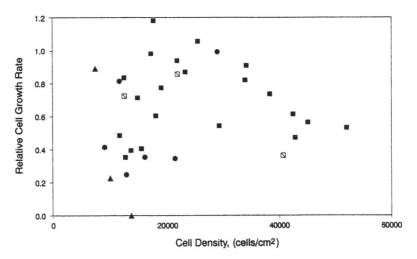

Figure 3.6 Plot of the relationship between cell density and relative cell growth rate for human dermal fibroblasts during three days of culture on TCP (□) or plasma surfaces toluene, *n*-hexane, MMA, EMA, methanol, ethanol, and isobutanol (□), FEP (▲) and plasma surfaces HMDSO or HMDSA (●).

Figure 3.7 shows the relationship between surface hydrophobicity, measured as the sessile contact angle and the density of cells attached. This figure contains data for the plasma surfaces toluene, *n*-hexane, MMA, HMDSO and HMDSA and data points for each of days 1, 2 and 3 of culture on each surface. It is clear that the cell densities achieved at days 1–3 of culture are at lower values on those surfaces with higher contact angles (that is, within the range 99–100 degrees) than on the more hydrophilic surfaces.

The ability of the plasma surfaces to support cell attachment and growth was also determined for human endothelial (HUE) cells (Figures 3.8 and 3.9). The surfaces used in these experiments were plasma films deposited on the metallized Kapton tape. As the serum adhesive glycoprotein fibronectin (Fn) has been shown to enhance endothelial cell attachment when adsorbed onto polymer surfaces prior to cell seeding, FEP coated with Fn and the effect of treating the plasma surface with Fn were determined as well. On this opaque surface, the number of cells attached was determined by fixation of the cell culture followed by staining with Eosin and visualisation of the cells using epifluorescence UV. Good HUE cell attachment was seen on the surfaces treated with toluene, triethoxyvinylsilane (TEVS), and MMA, whereas there was only poor attachment to the FEP surface. Pretreatment of the FEP with Fn markedly enhanced HUE cell attachment (Figure 3.8).

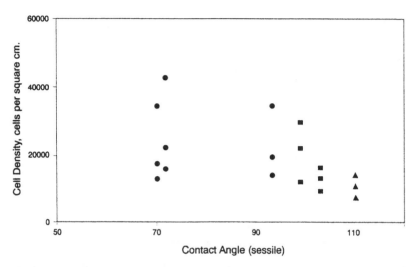

Figure 3.7 Relationship between the sessile contact angle of plasma modified surfaces toluene, *n*-hexane, MMA, HMDSO, HMDSA and unmodified FEP, and the density of human dermal fibroblasts after one, two and three days of culture.

The HMDSO plasma surface was clearly enhanced over the unmodified FEP surface for the ability to attach HUE cells, but the attached cells did not form the same well-spread morphology that was observed on the other plasma surfaces. Figure 3.9 shows the effect of culture time on HUE cells cultured on an MMA plasma surface, on an MMA surface that was pre-coated with Fn, and on FEP precoated with Fn. The density and spreading of the HUE cells on the MMA surface was sufficiently good for there to be little additional stimulatory effect on both cell morphology and growth of precoating with Fn. On FEP coated with Fn, initial cell attachment and growth was quite good, but lifting off of the cell monolayer from the substratum after several days' culture was observed in some cultures.

The present results document pronounced effects on cell attachment and growth by chemical modification of polymer surfaces. These studies suggest that it may ultimately be possible to design the chemical composition of a surface for optimum attachment and growth of cells and other biomedical performance. However, at present the mechanistic understanding of the phenomena at the biological/surface interface are very incomplete. It is reasonable to surmise that the above cell growth effects are due to different adsorption behaviour of proteins onto the polymer surfaces under study. Biological systems of interest invariably contain complex mixtures of proteins. The relative efficiency of the various proteins competing for surface adsorption sites, the denaturation processes of adsorbed proteins,

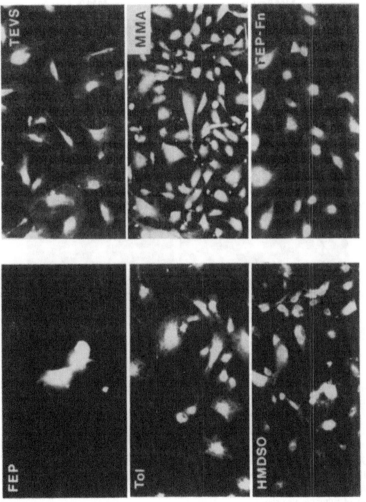

Figure 3.8 Cell attachment and morphology of human endothelial cells cultured for three days on FEP and on toluene, HMDSO, TEVS, and MMA plasma surfaces and on FEP, which had been precoated with fibronectin (FEP-Fn).

77

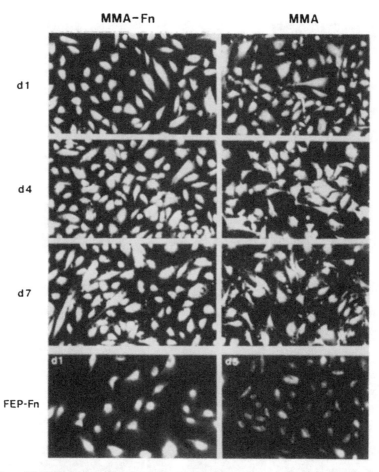

Figure 3.9 Human endothelial cell attachment and spreading after one, four, or seven days of culture on MMA plasma surface, and on MMA surface, which had been precoated with Fn prior to seeding with the cells. The attachment and spreading of cells grown on Fn-coated FEP after one or five days of culture is also shown.

and the effects of the composition of the adsorbed protein layer on cell growth are topics for further studies, which may lead to an understanding of the above phenomena. In conjunction with work on selective surface modification and improved polymer surface analysis, this may ultimately lead to the ability to design the interface between polymers and biological environments for optimal performance in various applications (which differ by the proteins present) and enable the rational design of new products on the basis of surface engineering. For the design of polymer surfaces with controlled biological responses, however, much more work is needed on the detailed characterization of the groups present and their density on the outermost surface layers of plasma polymers that give the best responses.

3.8 REFERENCES

1 Yasuda, H., Sharma, A. K,. Yasuda, T. *J. Polym. Sci.: Polym. Phys. Ed.*, 19:1285, 1981.

2 Ratner, B. D., Johnston, A. B. and Lenk, T. J. *J. Biomed. Mater. Res.: Appl. Biomater.*, 21:59, 1987.

3 Andrade, J. D. In: *Surface and Interfacial Aspects of Biomedical Polymers, Vol. 2*, J. D. Andrade, ed., Plenum Press, P. 1, 1985.

4 Gerenser, L. J., Elman, J. F., Mason, M. G. and Pochan, J. M. *Polymer*, 26:1162, 1985.

5 Hollahan, J. R. and Bell, A. T., eds., *Techniques and Applications of Plasma Chemistry*. Wiley, New York, 1974.

6 Coopes, I. H. and Gifkins, K. J. *J. Macromol. Sci.-Chem.*, A17:217, 1982.

7 Winters, H. F., Chang, R. P. H., Mogab, C. J., Evans, J., Thornton, J. A. and Yasuda, H. *Mater. Sci. Eng.*, 70:53, 1985.

8 Yasuda, H. *Plasma Polymerization*. Academic Press, Orlando, FL, 1985.

9 Yasuda, H., ed. *Plasma Polymerization and Plasma Treatment of Polymers*, Wiley, New York, *J. Appl. Polym. Sci.: Appl. Polym. Symp.*, 42, 1988.

10 Gerenser, L. J. *J. Adhesion Sci. Tech.*, 1:303, 1987.

11 Collins, G. C. S., Lowe, A. C. and Nicholas, D. *Europ. Polym. J.*, 9:1173, 1973.

12 Yasuda, H., Marsh, H. C., Brandt, S. and Reilly, C. N. *J. Polym. Sci.: Polym. Chem. Ed.*, 15:991, 1977.

13 Griesser, H. J., Hodgkin, J. H. and Schmidt, R. In *Progress in Biomedical Polymers*, C. G. Gebelein and R. L. Dunn, eds., Plenum Press, p. 205, 1990.

14 Griesser, H. J. *Vacuum*, 39:485, 1989.

15 Jaffe, E. A., Hoyer, L. W. and Nachman, R. L. *J. Clin. Invest.*, 52:2757, 1973.

16 Youxian, Da, Griesser, H. J., Mau, A. W. H., Schmidt, R. and Liesegang, J. *Polymer*, in press (1991).

17 Clark, D. T. and Dilks, A. *J. Polym. Sci.: Polym. Chem. Ed.*, 17:957, 1979.

18 Griesser, H. J., Youxian, Da, Hughes, A. E., Gengenbach, T. R. and Mau, A. W. H. *Langmuir*, submitted.

19 Griesser, H. J., Meijs, G. F. and McAuslan, B. R. *J. Bioact. Compat. Polym.*, 5:179, 1990.

20 Ratner, B. D., Johnston, A. B. and Lenk, T. J. *J. Biomed. Mater. Res.: Appl. Biomater.*, 21:59, 1987.

21 Briggs, D. In *Polymer Surfaces and Interfaces,* W. J. Feast and H. S. Munro, eds., Wiley, New York, p. 33, 1987.

22 Chilkoti, A. Lyons, C. S. Personal communications.

23 Chatelier, R. C., Gengenbach, T. R., Vasic, Z. R. and Griesser, H. J. Paper presented at the *Conference on Surface Engineering*, Adelaide, South Australia, 1991.

24 Coopes, I. H. and Griesser, H. J. *J. Appl. Polym. Sci.*, 37:3414, 1989.

25 Griesser, H. J. and Chatelier, R. C. *J. Appl. Polym. Sci.: Appl. Polym. Symp.*, 46:361, 1990.

26 Yu, D. and Marchant, R. E. *Macromol.*, 22:2957, 1989.

27 Sakata, J. and Wada, M. *J. Appl. Polym. Sci.*, 35:875, 1988.

28 Gengenbach, T. R., Vasic, Z. R. and Griesser, H. J. Unpublished results.

CHAPTER 4

The Present and Future for Coil-Coated Products in the Building Industry

R. A. SIMCOCK[1]

4.1 INTRODUCTION

ORGANIC coil-coated steel products have, over the last twenty years, enjoyed rapid growth rates, and this trend is continuing. Figure 4.1 shows the growth rate in Australia since 1971; the current production rates for domestic and export markets are approximately 300,000 tonnes per annum. Statistics from the European Coil Coaters Association and the National Coil Coaters Association show similar trends for Western Europe and North America, respectively.

The available organic coatings range from paint systems formulated to provide coloured, durable surfaces to a number of laminated films that provide the additional properties of texture and toughness. Metal/polymer/metal laminates are used to fabricate sound-deadening materials, and a significant market in the United States is their use to prevent squealing of disc brakes.

The building industry in Australia forms the largest single market segment for prepainted steel (Figure 4.2). Prepainted steel offers the building industry large coloured surfaces with structural strength and integrity at a low cost. Large buildings can be clad quickly and efficiently. The development of novel metallic coated steels and paint systems, combined with an efficient painting process, has enabled products of excellent durability to be offered by BHP Steel (on its COLORBOND®* range) with warranties

*Trademark of John Lysaght (Australia) Limited.

[1]BHP Steel-Coated Products Division, Research & Technology Centre, P.O. Box 77, Port Kembla, NSW 2505, Australia.

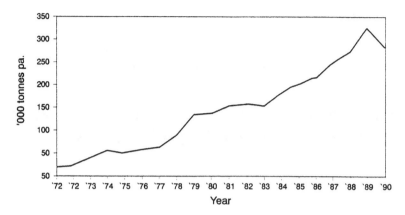

Figure 4.1 COLORBOND® prepainted steel production.

of fifteen to twenty years, depending on the paint system and environment of the installation. In recent years, prepainted steel has been used to clad numerous prestige buildings, particularly those related to the Australian Bicentennial celebrations. Darling Harbour in Sydney and the foam-filled sandwich panels used extensively at Expo 88 are two good examples of prepainted steel being used to add style and colour to a project.

4.2 THE PROCESS

Coil coating is an efficient painting process that provides thorough cleaning, complete pretreatment, and consistent paint application to the

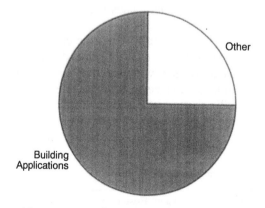

Figure 4.2 COLORBOND® prepainted steel sales by market, 1989/90.

steel substrate. As the process is carried out on steel coils, the strip is flat, allowing complete access to the total surface area for cleaning, pretreatment, and paint application. The basic coil coating process is illustrated in Figure 4.3. The strip is cleaned with alkaline solutions by either an immersion or spray application to remove oils and other surface contaminants. Following cleaning, the sheet is rinsed, then chemically pretreated to prepare the surface for priming. The type of pretreatment depends upon the substrate being painted; typically, galvanized steel has a zinc phosphate or mixed oxide pretreatment, whereas ZINCALUME®* coated steel and aluminum have chromate pretreatments. After pretreating and drying, the strip passes to the primer coater where the primer is applied to both sides of the strip. The paint is applied using roll coater technology to ensure uniform wet film thickness. The wet primed strip passes into the primer oven where it is heated to cure the primer. After quenching, the finish coat is applied to the primed strip, cured in the main oven, quenched, and recoiled.

Laminates are produced by applying the adhesive in the top coater followed by lamination of the film as the hot strip emerges from the main oven. Protective plastic films can also be applied to the painted strip at the same location.

To allow continuous coating of steel strip, entry and exit accumulators coupled with stitching facilities allow coils to be joined together successively without stopping the constant speed section of the line. Shearing after the exit accumulator allows coils of the mass required by the end customer to be produced.

The main advantage of prepainting is that since the strip is flat, all parts of the coil surface can easily be reached for cleaning, pretreatment, and painting. Control of the nine stages of pretreatment and film thickness of both primer and finish coat can be accurately and continuously monitored. Environmentally, the process can be considered friendly, as spent solutions from the pretreatment section are treated before discharge. The volume of water used is kept to a minimum by methods such as cascading of rinse waters. Fumes generated from solvent-based paints are taken from the coater room into the ovens, and the oven exhaust is incinerated before discharging.

4.3 THE PRODUCT

The typical prepainted product supplied to the building industry consists of a pretreated substrate, 5 μm of primer and 20 μm of finish coat. All prepainted steel must have adhesion of the highest level to withstand subsequent fabrication into the finished product. Typically, for a roll-formed

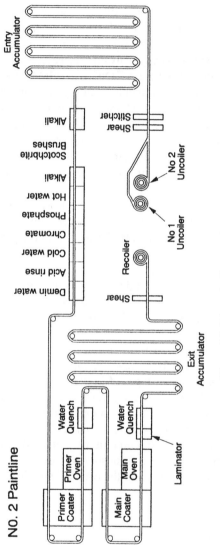

Figure 4.3 Schematic diagram of a paint line.

84

prepainted steel roofing sheet, the level of paint adhesion should be such that the prepainted steel can be bent around a diameter equal to two to three times metal thickness (2–3T). In specific end uses where maximum flexibility is required, then the paint can be formulated so the adhesion is retained even though the prepainted steel is bent flat on itself.

As the major outlet for prepainted steel is the building industry, the corrosion resistance of the substrate and the durability of the paint system become two important characteristics. The majority of the Australian population lives around the coast, making it essential for any building product to have good corrosion resistance. Equally, however, the level of ultraviolet radiation to which Australia is exposed is significantly greater than the larger markets of North America, Europe, and Japan. These two demands on the performance of prepainted steel by the Australian building market have initiated significant developments by BHP together with its paint suppliers to achieve today's product.

4.3.1 CORROSION RESISTANCE

Galvanized steel is the substrate most commonly painted in the United States and Europe. Much of the Australian market, however, has a marine environment where galvanized steel frequently gives rise to unexpected and unacceptable levels of corrosion. The introduction, by BHP, of ZINC-ALUME (55% Al–45% Zn) coated steel, originally developed by Bethlehem Steel in the U.S.A., resulted in a prepainted steel with significantly improved corrosion resistance, particularly in marine environments.

The superior performance in salty conditions of unpainted ZINC-ALUME coated steel compared to unpainted galvanized steel is illustrated by the results from salt spray testing (ASTM B117–64) reported by Harvey [1] who obtained a corrosion rate of 13–19 h/g for passivated ZINC-ALUME and 2.3 h/g for passivated galvanized steel [2].

This better corrosion resistance of ZINCALUME is also found on atmospheric exposure. Figure 4.4 shows the corrosion rates of galvanized steel and ZINCALUME steel at a severe marine site [1]. A similar series of exposures, conducted at other test sites in different environments, confirmed the better corrosion resistance of ZINCALUME when compared to galvanized steel although their relativity varies with the aggressiveness of the test site [1]. The results are given in Table 4.1, and they can be used to calculate expected life of commercial products at each site (Table 4.2).

Prepainted ZINCALUME usually exhibits better corrosion resistance in salt spray testing compared with prepainted galvanized steel, although the ZINCALUME-based material tends to corrode more rapidly from unprotected sheared edges. The amount of edge corrosion of prepainted ZINCALUME can be reduced significantly by using specially formulated primers.

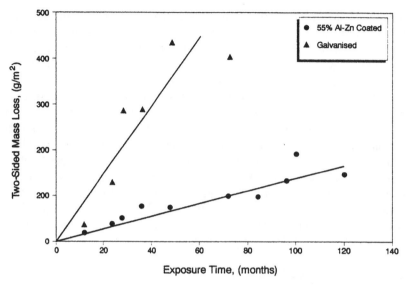

Figure 4.4 Corrosion rates of galvanized steel and ZINCALUME at severe marine site.

On natural exposure at the severe marine test site, the corrosion resistance of the prepainted ZINCALUME is better than prepainted galvanized steel. The main improvement is related to the formation of white corrosion product, the prepainted galvanized product generating more than the equivalent ZINCALUME product. One surprising result from this natural exposure testing was the small difference in the corrosion resistance of prepainted ZINCALUME when primed with the original standard epoxy primer and the primer reformulated to improve the performance of prepainted ZINCALUME in salt spray testing.

TABLE 4.1. Corrosion Rates of Galvanized Steel and ZINCALUME Coated
Steel at Australian Atmospheric Exposure Sites.

Site	Galvanized Steel $g/m^2/yr$	ZINCALUME Steel $g/m^2/yr$
Bellambi Point, NSW (severe marine)	140	16
Blacksmiths, NSW (marine)	18	4
Port Kembla, NSW (industrial/marine)	20	4.2
Stroud, NSW (rural)	4	1.3

TABLE 4.2. Estimated Lives of 450 g/m² Galvanized and 150 g/m² ZINCALUME Coated Steels at Australian Test Sites.

Site	Galvanized Steel 450 g/m²	ZINCALUME Steel 150 g/m²
Bellambi Point, NSW (severe marine)	3*	9*
Blacksmiths, NSW (marine)	25	38
Port Kembla, NSW (industrial/marine)	23	36
Stroud, NSW (rural)	113	100

*Because of the severe nature of this site, edge effects resulted in an overestimate of the corrosion rate. Actual samples of galvanized steel perforated in about five years and ZINCALUME in about eighteen years.

This better corrosion resistance of the prepainted ZINCALUME, compared with prepainted galvanized steel, is especially evident when both products are exposed in areas of buildings not washed by rain, for example, under eaves. Figure 4.5 shows the two prepainted products exposed adjacent to each other under the eaves of a building. The absence of the white corrosion product on the prepainted ZINCALUME substrate (left-hand panel in Figure 4.5) is a measure of its superior corrosion resistance.

Figure 4.5 Performance of prepainted galvanized and ZINCALUME in a sheltered area.

This improved corrosion resistance has since been confirmed many times in many different locations.

A further improvement in the corrosion resistance of prepainted ZINC-ALUME can be achieved by increasing the primer thickness and changing the primer composition. Increasing the primer thickness from 5 μm to 25 μm, changing its chemical composition, and increasing the corrosion inhibitive pigment content results in significantly increased corrosion resistance. A fluoropolymer finish coat to ensure excellent flexibility and colour retention completes the composition of the high build primer system. This product, COLORBOND XSE, can be recommended confidently in all but the most severe marine or industrial environments and shows improved life.

4.3.2 PAINT DURABILITY

The three paint systems most commonly used in Australia on exterior building products are polyester (PE), silicone-modified polyester (SMP), and polyvinylidene (PVDF). In the U.S.A., Europe, and Japan, PVC plastisol is often used; however, in the Australian environment with its high UV radiation, the stability of PVC plastisols is doubtful. Samples exposed at BHP's test sites have shown very unpredictable performance and, therefore, have never been introduced into the COLORBOND prepainted steel range.

As the prepainted cladding of buildings receives little, if any, maintenance, e.g., cleaning or washing to retain the original appearance, then the stability of the resin and pigments is of major importance. It is incorrect to assume that the performance of all polyesters or SMP is the same. Very early, Australian exposure data of paint formulations developed for U.S. or European conditions showed that extrapolation of European or American results to Australian conditions was extremely risky. These early results led BHP Steel to avoid the use of organic pigments in paint formulations and only use the SMP paint system for exterior use in the building market.

Figure 4.6 shows the change in colour and the rate of chalk formation of a blue polyester paint system available in the late 1970s, after three years' exposure at Rockhampton. (ΔE measures colour change; $\Delta E = 2$ is just noticeable to the unaided eye; $\Delta E = 10$ easily noted colour change.) The cause of the colour change is twofold, the fading of the blue organic pigment and the poor UV stability of the polyester resin, which is reflected in a high rate of chalk formation. The same poor pigmentation in a more chalk-resistant SMP paint system shows better chalk resistance, but the loss of colour is still unacceptable (Figure 4.7). Similarly, Figure 4.8 is a light yellow colour with most fading being caused by chalking of the resin rather than loss of colour of the pigment. Of particular interest was a dark brown colour (Figure 4.9), which contained only stable inorganic pig-

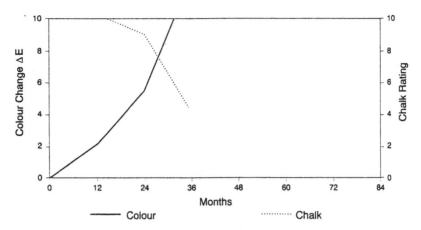

Figure 4.6 Polyester blue colour (87052-04) – Rockhampton.

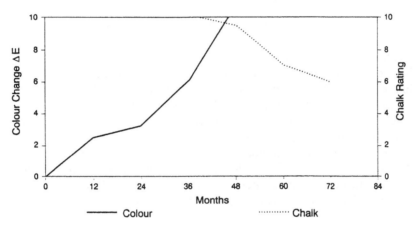

Figure 4.7 SMP blue (84049-2C) – Rockhampton.

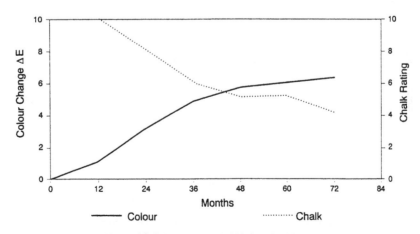

Figure 4.8 Polyester gold (84020-0l) – Rockhampton.

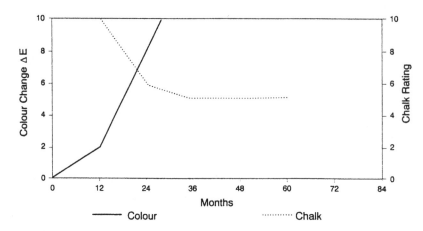

Figure 4.9 Polyester dark brown (84171-03) — Rockhampton.

ments such as iron oxides and TiO_2. The chalk formation is still excessive, but the very rapid loss of colour, even when the same pigmentation was put into the more chalk-resistant SMP resin system (Figure 4.10), led us to consider other mechanisms by which this paint system could fade.

Analytical examination of the surface of the weathered paint film by X-ray diffraction and electron microscopy [3] demonstrated that the main cause of colour change was not instability of the pigments, but a change on the surface of the paint film in the ratio of the four stable component pigments of this colour. It was demonstrated that very small pigments were preferentially lost by erosion from the surface of a weathered paint film,

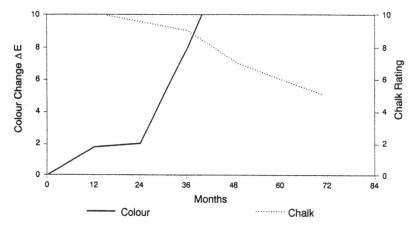

Figure 4.10 SMP dark brown (84071-3C) — Rockhampton.

and long acicular pigments were, because of their shape, preferentially retained on the paint film surface. For long-term colour stability of a weathering paint film, it is necessary to select pigments of similar shape and size so that the ratio of these pigments in the initial surface is retained as the films weathered. In those instances where this requirement cannot be satisfied because of the pigments needed to produce the required colour, the colour will probably change as the paint film weathers.

If the UV resistance of the polyester resin is poor, then even with pigments of the same size and shape referred to earlier, these paints will chalk and lose colour because white opaque chalk forms on the paint surface. Since the early 1980s, the durability of the available PE resin system has improved significantly, to the point where some PE paint systems used today have durability equivalent to SMP systems.

Figure 4.11 is a light yellow polyester showing excellent chalk resistance and colour retention; similarly for the dark brown (Figure 4.12) and blue (Figure 4.13) colours. These are typical of the performance of the polyester paint systems available today, although paint systems of this quality are not always used internationally.

Fluorocarbon resin-based paint systems, PVDF, have long been recognised and accepted as the paint systems with the most chalk-resistant resin and colour retentive properties of all available paint systems. For comparison, data of their exposure performance is given in Figures 4.14 and 4.15.

BHP's COLORBOND steels generally have a ZINCALUME substrate, and the paint system can be expected to retain its integrity for ten, fifteen or twenty years, depending upon the paint system applied to the ZINCALUME. The product will have been manufactured on a continuous paint

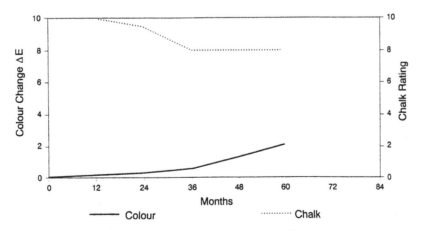

Figure 4.11 Polyester gold (85026-03) – Rockhampton.

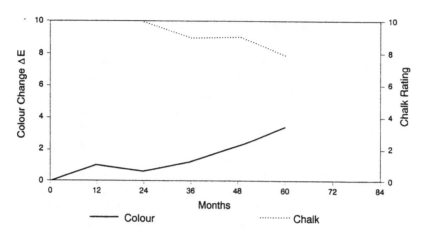

Figure 4.12 Polyester dark brown (85026-06) – Rockhampton.

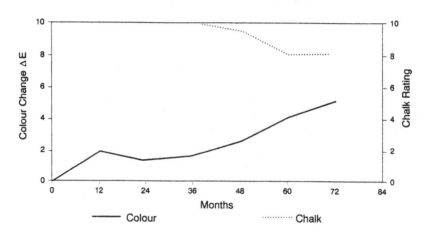

Figure 4.13 Polyester blue (84136-01) – Rockhampton.

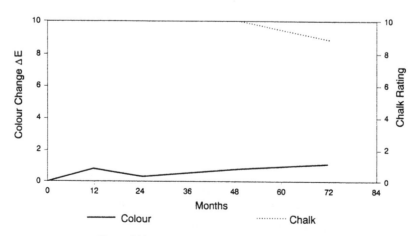

Figure 4.14 PVDF gold (84046-5C) – Rockhampton.

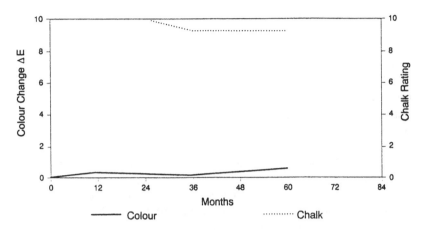

Figure 4.15 PVDF dark brown (85004-02) – Rockhampton.

line operated with a very high degree of control with all liquid effluent treated before discharge, and the solvents generated during the heat curing of the paint systems will have been incinerated.

4.4 THE FUTURE

Looking to changes to be expected in the coil coating industry in the 1990s, not only have shortcomings of existing products and processes to be considered, but also outside influences that will be superimposed on the industry. It is this latter consideration, in the form of safety and environmental factors, which may bring about the most significant changes to the product and the process in the 1990s.

The durability and corrosion resistance of the range of prepainted steel products available today, particularly if prepainted stainless steel is included, satisfy the majority of applications. There will always be those specific end uses that will require separate consideration. There is a general need to improve the damage resistance of these prepainted steels with a total paint film thickness of 25 μm. Damage to the paint film, which occurs during roll forming, transportation, and installation is, at present, unacceptably high. The current solution to this problem is to apply a 75-μm polyethylene film to the painted surface at the paint line. The film protects the painted surface during fabrication and transportation of the product and is removed immediately after installation. This procedure generates large amounts of waste plastic film on the construction site, and occasionally, the plastic film cannot be removed from the paint film because of the development of a permanent bond to the painted surface.

Both problems could be avoided by the development of a more damage-resistant paint system.

The use of polyamide beads in the paint formulation is widely and successfully used in Europe on aluminium for products such as security blinds, where the continuous rubbing of adjacent slats removes a conventional paint system. These paint systems on steel substrates also give an improvement in damage resistance (Figure 4.16), but for their use as a roofing or walling material, the effect of sunshine on the beads/paint needs to be established, and the consequences of the reduction in coefficient of friction brought about by the presence of the beads needs to be examined.

Solvents used in paints constitute a vast research area for future developments. Over the last few years, particularly offensive solvents have been replaced in paint formulations as part of a continuing product development programme. In making process or paint formulation changes related to solvents, how a paint line handles solvents needs to be understood. Typically, in BHP Steel paint lines, the coater rooms, where paint is applied to the strip, are enclosed areas. Painting by its nature will generate some solvent emissions. The solvent-containing air from the coater room is continuously extracted into the curing ovens to be mixed with the solvents released during the curing of the paint film. In turn, the ovens are continuously exhausted into an incinerator operating at 760°C. The energy generated is then used for the heating of the pretreatment section of the paint line and the building itself. Because of this process, the use of high volume solids paint systems, e.g., >70% volume solids, will have little effect on the process since incineration will still be required.

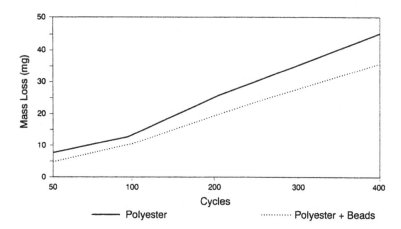

Figure 4.16 Effect of polyamide beads on the taber abrasion resistance of a polyester paint system.

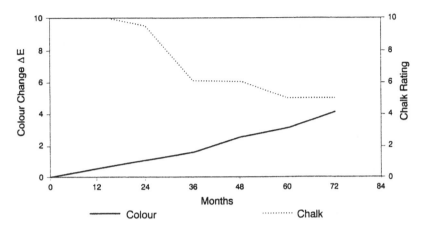

Figure 4.17 Water-based acrylic gold (84013-12) – Rockhampton.

Water-based paint systems have been available for some years, but even they are not free of solvents, as up to 15% solvents can be included in their formulation. The durability of water-based paint systems to high UV radiation is good, as shown in Figure 4.17, which shows chalk and colour change of a water-based paint system when exposed in Rockhampton, but their corrosion resistance has to be established.

If solvent-free or 100% solids coatings are to be developed, there are two considerations: how to apply them and how to cure them. Powder coating is an area of painting that has grown significantly in recent years [4]. It is an efficient process, especially when overspray is collected and reused, with few of the environmental problems of wet painting. Perhaps powder coating needs to be considered as a means of coil coating and developed further. The application methods will need to be refined if total film thicknesses of 20–25 μm are to be consistently achieved on steel strip. The quality of the surface finish of thin powder coatings is poor compared to a coil-coated finish, and the ability to change colour quickly is a significant challenge (the colour range of prepainted steel can be several different colours). The final consideration is that of long-term durability. The colour and chalk resistance of current powder coatings is short of that attainable with existing coil-coated finishes (Figure 4.18), so for the immediate future, these coatings might be confined to interior applications where UV radiation is low, e.g., domestic appliances.

Longer term work on 100% solids coatings must consider electron beam (EB) curing of the paint system. There are problems to be resolved before a commercial product is available. The main ones are the generation of ozone and X-rays, which will require exhaust and shielding facilities and

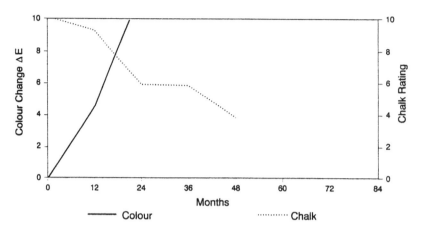

Figure 4.18 Powder coating dark brown (86004-15) – Rockhampton.

the smell/toxicity of some monomers and reactive diluents. Initial coil coating experiences have indicated difficulty in achieving adhesion to metal substrates.

However, the benefits to be gained from an EB-cured system in coil coating are very significant. Firstly, there is the possibility of formulating paint films with properties that are currently unattainable, particularly very hard paint films with exceptional flexibility. The cure times for EB systems are extremely short, which promotes the possibility of painting at high line speeds, e.g., 200–300 m/min. High-speed painting, however, may not be the major benefit of short cure times. The benefits of rapid curing will be reflected in the design of a paint line that can be stopped and started without loss of the product in the line.

This ability would significantly change the design of a paint line, eliminating the need for accumulators, constant speed sections, quick colour change facilities, and long ovens to cure the paints, which, consequently, makes them cheaper to install. Control of the process would require the use of sophisticated control mechanisms that may need to be developed. For example, continuous accurate monitoring of paint film thickness will be one of the parameters needed to control the economics of the process.

Although there have been improvements in the performance of polyester paint systems in recent years, further improvements can be expected in the future. Today, the curing of polyester systems is usually by reaction with melamine derivatives or blocked isocyanates. This cured resin is vulnerable to hydrolytic attack at the crosslink site. The recent availability in commercial quantities of novel crosslinkers, e.g., acetoacetate derivatives [5], and novel cyclic aliphatic diacids as monomers for the synthesis of the polyesters should lead to the development of interesting products, which may have better hydrolytic stability.

A new resin manufactured by the copolymerization of vinyl ethers and polyvinylidene difluoride (FEVE) is reported to have similar weathering properties to the PVDF [6] resin. Advantages claimed for the FEVE resin are the ability to formulate a high-gloss paint system, the presence of reactive groups allowing crosslinking to take place more easily, resulting in better adhesive properties. Paint systems based on FEVE resins are reported to have higher volume solids and better mechanical propeties. To the coil coater, these properties translate into easier paint application and, consequently, greater production speeds.

It is the coil coating process that will probably undergo the greatest changes in the 1990s. The pretreatment section of the paint line will need to operate without liquid discharge. This will involove the development of a process to clean and pretreat the steel strip very efficiently with low volumes of solutions and then a recovery system to allow the recovered water to be reused and the extracted solid materials to be returned to the supplier for reworking into the original product. Work in this direction has started at BHP's new No. 5 Paint Line at Brisbane where the water for the painting process is rainwater collected from the roof of the building housing the paint line. New developments in reverse osmosis and ion exchange technology will be used to treat effluent waters before they are recycled.

Current spray or dip processes to pretreat strip will be replaced by dried-in-place systems. With these systems, the correct level of pretreatment is metered onto the clean strip, then dried. This is unlike most spray or drip systems, which require excess chemicals to be rinsed away, giving rise to large volumes of rinse waters that need treatment before discharge.

4.5 SUMMARY

In summary, the 1990s will see significant changes in the process of coil coating, yet the quality of the prepainted product will continue to improve. Changes to reduce or eliminate the need for organic solvents in paint systems will eventually lead, in the longer term, to new chemistry being used to produce the resins. In the shorter term, novel chemical monomers and intermediates should allow for the continued improvement in properties of organic coatings for steel, but most of these improvements will be related to the process and properties outside the more usual area of colour and corrosion resistance.

4.6 REFERENCES

1 Harvey, G. J. "Properties and Uses of Bare and Prepainted 55% Al-Zn Alloy Coated Steel," *Revue de M' etallurgie* (2):183, 1990.

2 Harvey, G. J. and Richards, P. N. "Zinc-Based Coatings for Corrosion Protection of Steel Sheet and Strip," *Metals Forum*, 6(4):234–247, 1984.

3 Boge, E. M., Mercer, P. D. and Simcock, R. A. "The Exterior Durability of Paint Systems for Coil Coatings," *Surface Coatings* (July):6–10, 1986.

4 Jotischky, H. Paint Research Assoc., "Powder Coatings in the Market-Place," in *Thermoset Powder Coatings*, J. Ward, ed., Redhill, Surrey, pp. 134–7, 1989.

5 Del Rector, F., Blount, W. W. and Leonard, D. R. "Applications for Acetoacetyl Chemistry in Thermoset Coatings," *Journal of Coatings Technology,* 61(771): 31–37.

6 Munekata, S. "Fluoropolymers as Coating Materials," *Progress in Organic Coatings,* 16:113–134, 1988.

CHAPTER 5

Metallurgical and Optical Thin Films Produced by Filtered Arc Evaporation

P. J. MARTIN[1]
R. P. NETTERFIELD[1]
T. J. KINDER[1]
L. DESCÔTES[2]

5.1 INTRODUCTION

THE vacuum arc has been used in the deposition of metallurgical coatings for the last twenty years. Its origin as a deposition technology can be traced to the early work of Soviet researchers, and there have been several commercial coating machines developed from the early patents of Sablev [1] and Snaper [2]. The dc arc sources have been satisfactory for the deposition of a wide range of materials, but most attention has been focussed on TiN for wear-resistant applications on tools. There has also been some activity in the development of pulsed arcs for high rate deposition of surface alloys [3].

The attraction of vacuum arcs in coating technology lies with the relative simplicity of the technology and, more importantly, the high degree of ionisation in the emission. The ionisation and intrinsic energy of the emitted ions leads to increased reactivity of the metal vapour with reactive gases for compound formation and increased hardness and adhesion of the deposited layers. The major problem of the arc process is the presence of large micron-sized droplets of cathode material in the vapour stream. These *macroparticles* have restricted the application of the arc process to the metallurgical coating of machine tools and wear parts. This chapter describes a method of removing these macroparticles and the deposition of metallurgical and optical coatings.

[1]CSIRO, Division of Applied Physics, Sydney 2070, Australia.
[2]Thomson CSF, Paris, France.

5.2 CATHODIC ARC EMISSION

Figure 5.1 shows a schematic of the arc evaporation process. The cathode spot of diameter 1–20 microns is a source of electrons, metal atoms, and macroparticles. Electron-atom collisions lead to the formation of ions, which are accelerated toward the cathode and become possible sources for new emission sites. Ions are also ejected from the positive ion cloud immediately above the cathode spot toward the anode. Although the precise nature of the emission is not well understood, there is general agreement that the potential hump resulting from the ion cloud accounts for the high energies of the emitted ion, which are typically 50 volts above cathode potential. The macroparticles are emitted at an angle of approximately 10–20 degrees to the cathode plane, and there have been several reports of filtering based upon the use of optical baffles. However, the most successful approach to filtering has been to use an Aksenov [4] plasma duct to magnetically steer the plasma into a deposition chamber.

5.3 FILTERED ARC SYSTEM

The problem of plasma flux motion along a toroidal magnetic field has been extensively studied [5]. The filtering effect is derived from the combination of a low strength magnetic field of around 100 Gauss and the elec-

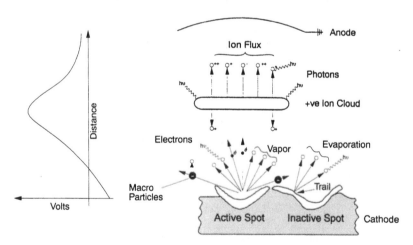

Figure 5.1 Schematic of the basic arc evaporation processes and the potential distribution above the cathodic spot.

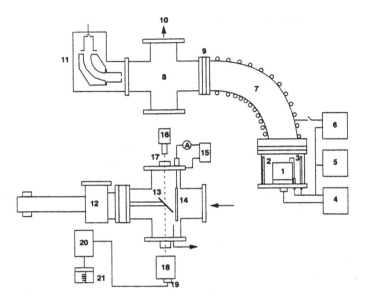

Figure 5.2 Schematic diagram of a filtered arc system showing the deposition chamber fitted with (a) optical film diagnostics and (b) beam diagnostics: 1. cathode, 2. anode, 3. trigger, 4. supply, 5. supply 6. bias supply, 7. plasma duct, 8. chamber, 9. isolator, 10. pump, 11. ion-energy analyser, 12. load-lock, 13. substrate, 14. shutter, 15. bias supply, 16. lamp, 17. window, 18. monochromator, 19. photodiode, 20. amplifier, 21. chart recorder.

tric field due to the electrons in the plasma. Figure 5.2 shows a schematic diagram of the essential features of the CSIRO filtered arc system.

The cathode is 58 mm diameter and mounted on a water-cooled copper housing. The anode is electrically isolated from the housing and also water-cooled. The arc is ignited by a pneumatically controlled tungsten trigger wire and activated by an automatic control system. The beam is focussed and transported by magnetic fields around the plasma duct to the deposition chamber. The chamber can be configured for beam character-isation or film deposition. In the latter case, the chamber is equipped with optical monitoring for oxide deposition and fitted with a load-lock for rapid sample introduction. The transmitted beam is found to have a f.w.h.m of 30 mm and ion currents up to 1 A were measured.

Conducting substrates could be heated by electron bombardment from the plasma by positively biasing the substrate. In the case of glass sub-strates, a quartz lamp was used for radiative heating. The temperature was monitored by a thermocouple. Reactive deposition was achieved by intro-ducing reactive gases into the deposition chamber immediately in front of the substrate assembly through a gas flow control system. Substrates used are conducting silicon wafers (0.05 Ω cm) and microslides.

5.4 RESULTS AND DISCUSSION

The effectiveness of the filtering was assessed by measuring the deposition rate with and without the magnetic field in the duct switched on and by electron microscopy of the films. The deposition rate for a 90 A Ti arc was found to be 40 μm hr^{-1}. Under no-field conditions, the rate was 30 nm hr^{-1} and assumed to result from Ti reflected around the walls of the duct. No macroparticles were observed in either case.

The energy distribution of the beams transported to the energy analyser from Ti and C cathodes is shown in Figure 5.3. The peak energies were found to be 44 eV for Ti and 28 eV for C. These values are consistent with previous measurements of Davis and Miller [6]. The intensity of Ti energy distribution was reduced as the pressure of nitrogen in the deposition chamber was raised, and some evidence of a slight shift to lower energies was observed at 5×10^{-4} torr. This shift is due to energy loss in the Ti-N$_2$ gas phase collisions.

The microhardness of a range of materials deposited by the filtered arc is given in Table 5.1. The optical properties are given in Table 5.2. The hardness values are consistent with those reported for conventional arc deposition. The optical properties are equivalent to the best values reported for ion-assisted thin film deposition.

The deposition rates for Al$_2$O$_3$ and ZrO$_2$ were found to be 35 μm hr^{-1} and 20 μm hr^{-1}, respectively. The refractive index for ZrO$_2$ was calculated

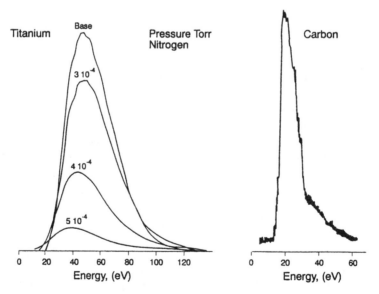

Figure 5.3 Ion energy analysis for titanium and carbon beams.

TABLE 5.1. Microhardness Values for Arc-Deposited Films.

Film	Microhardness (H_{V5})
TiN	2550
ZrN	2780
C	3600
C + CH$_4$	1600
ZrO$_2$	1040
Al$_2$O$_3$	1080

from the maxima and minima in the transmittance monitoring trace at 633 nm. A value of 2.19 was calculated, which compares with 2.148 for the bulk value and 2.19 for ion-plated films [7]. High-quality Al$_2$O$_3$ films with an index of 1.67 were also deposited. Films of AlN were also successfully deposited when the substrate temperature was raised to around 200°C in the presence of nitrogen, and a refractive index of 2.150 was measured. Stoichiometric VO$_2$ films were also deposited from a vanadium cathode at a substrate temperature of approximately 600°C. The film exhibited a semiconductor-metallic transition when heated above 68°C.

The films were found to be highly adherent and smooth. The r.m.s roughness of 0.8-μm thick film of Ti was found to be 0.8 nm compared to the silicon substrate value of 0.45 nm. Metal films were also successfully deposited onto kaptan, polyimide, and CR 39 plastic.

The stoichiometry of the TiN, TiO$_2$, and a TiC films was assessed by electron probe microanalysis and XPS. The binding energies of the Ti 2p

TABLE 5.2. Optical Properties of Arc-Deposited Films.

Film	Index of Refraction (n_{633})	Absorptivity
C	2.490	0.71
Graphite	2.030	0.70
C + CH$_4$	1.810	0.02
ZrO$_2$	2.190	<0.0015
Al$_2$O$_3$	1.670	<0.0005
AlN	2.150	<0.0010
VO$_2$ (ambient)	3.56	1.16
	2.56*	0.14*
VO$_2$ (68°C)	2.96	0.41
	1.57*	1.72*

*At 1 μm.

peak in the deposited films were consistent with the bulk values. A film was deposited from a cathode of SAF 2205 stainless steel. The composition of this film was found to be the same as that of the cathode.

The structure of the carbon films was examined by electron energy loss measurements performed on samples deposited onto KCl and floated off in distilled water. Figure 5.4 shows the low-loss spectrum of the arc-deposited carbon compared to a film produced by glow discharge in methane. The peaks of the spectra correspond to the plasma resonance frequencies, which are determined by the density of valence electrons. The plasmon peak for diamond is around 33 eV, while for graphite it is around 27 eV [8]. The arc-deposited film has a peak at 32 eV, very close to the value for diamond. Further evidence for the diamond-like structure of the arc film is seen in the spectrum around 5 eV. The glow-discharge film shows a π feature that is characteristic of transitions between the π and π^* states [8]. These transitions show the presence of sp^2 or graphitic bonding, which is absent in the arc-deposited material.

The microhardness of the carbon films was also very high, indicating the diamond-like nature of the material. The material was, however, found to be highly stressed, and delamination from most substrates occurred after 24 hours. The film transparency was improved if the film was deposited in an atmosphere of hydrogen, the extinction coefficient reducing from 0.7 to 0.02. The film hardness was also reduced, but the adhesion improved.

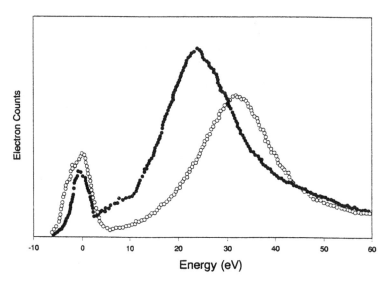

Figure 5.4 Electron energy loss spectra for glow-discharge carbon (●) and arc-deposited carbon (○).

5.5 CONCLUSION

The filtered arc method has been shown to be a practical high-rate deposition process for metals, nitrides, and oxides. The deposited films are free of macroparticles, smooth and adherent. The process should lend itself to most applications where metallurgical coatings are required and in optics and electronics.

5.6 REFERENCES

1 Sablev, L. P., Atamansky, N. P., Gorbunov, V. N. and Dolotov, J. I. U. S. Patent No. 3,797,179, 1974.

2 Snaper, A. A. U. S. Patent No. 3,625,848, 1971.

3 Boxman, R. L. "Fast Deposition of Metallurgical Coatings and Production of Surface Alloys Using a Pulsed High Current Vacuum Arc," *Thin Solid Films,* 139:41–52, 1986.

4 Aksenov, I. I., Belokhvostikkov, A. N., Padalka, V. G., Repalov, N. S. and Khoroshikh, V. M. "Plasma Flux Motion in a Toroidal Plasma Guide," *Plasma Phys. Contr. Fusion,* 28:761–770, 1986.

5 Aksenov, I. I., Belous, V. A., Padalka, V. G. and Khorosikh, V. M. "Transport of Plasma Streams in a Curvilinear Plasma-Optics System," *Sov. J. Plasma Phys.,* 4:425–428, 1978.

6 Davis, W. D. and Miller, H. C. "Analysis of the Electrode Products Emitted by d.c. Arcs," *J. Appl. Phys.,* 16:2212–2221, 1969.

7 Guenther, K. H., Loo, B., Burns, D., Edgell, J., Windham, D., and Muller, K-H. "Microstructure Analysis of Thin Films Deposited by Reactive Evaporation and Reactive Ion Plating," *J. Vac. Sci. Technol.,* A7(3): 1436–1445, 1989.

8 Martin, P. J., Filipczuk, S. W., Netterfield, R. P., Field, J. S., Whitnall, D. F. and McKenzie, D. R. "Structure and Hardness of Diamond-Like Carbon Films Prepared by Arc Evaporation," *J. Mater. Sci.,* 7:410–412, 1988.

CHAPTER 6

Ion-Assisted Deposition: Film Properties and Process Monitoring

R. P. NETTERFIELD[1]
P. J. MARTIN[1]
T. J. KINDER[1]

6.1 INTRODUCTION

THERE are many material-related problems in vapour-deposited thin films. The physical properties of an evaporated film, such as adhesion, stress, hardness, and optical properties, have been shown to depend on process parameters, including substrate temperature, deposition rate, pressure, evaporation material, substrate material, and surface preparation. Since the process parameters are often difficult to control and adjust precisely, film properties can be difficult to reproduce, even in the same deposition system. Under optimum conditions, the film properties are usually inferior to those of the bulk material.

It has been shown in recent years that bombardment of the growing film with ions of energy in the range 50 to 1500 eV strongly influences the growth behaviour and therefore the resulting properties of the layer. One of the early findings was that of Hirsch and Varga [1] who showed that the adhesion of Ge films was increased and the stress decreased when bombarded with ions during growth. The term IAD is used to describe the deposition of films by evaporation with simultaneous bombardment of the growing film with ions generated in a separate source. The field has been the subject of many publications, several reviews [2], and a monograph [3]. Many of the process parameters in IAD can be varied independently so optimization of the film properties can involve many deposition cycles if film characterization is carried out *ex situ*. However, many *in situ* techniques are now available to provide real-time monitoring of certain properties of the film nondestructively. The merits of a range of optical tech-

[1]CSIRO, Division of Applied Physics, Sydney 2070, Australia.

niques have been recently reported [4], and the results obtained from two experimental set-ups are summarized here [5–6]. The use of *in situ* real time monitoring can greatly reduce the time it takes to optimize the deposition conditions. In this chapter, the use of ellipsometry and interferometry for *in situ* monitoring of optical properties, film thickness, deposition rate, and film stress will be emphasized.

6.2 EXPERIMENTAL

A full description of the equipment used to prepare the films and characterize them has been given elsewhere [5–6]. The essential features have been reproduced in Figure 6.1. The films are prepared by electron-beam evaporation onto unheated substrates. The growing film can be irradiated by inert or reactive gas ions from a Kaufman ion source [7]. The pressure in the systems is typically less than 10^{-5} Pa prior to deposition and usually operates at approximately 5×10^{-3} Pa during IAD film growth. Ellipsometric information was obtained at 633 nm at a fixed angle of incidence of 70 degrees relative to the substrate normal. Stress data were obtained from measurements of the curvature of a separate thin rectangular glass substrate using an interferometric system first described by Ennos [8].

6.3 RESULTS

Porous microstructure is one of the major problems with evaporated films. Since the arriving vapour species at the substrate have very little energy (of the order of 0.1 eV), they have low surface impact mobility. The resulting columnar microstructure is due to atomic self-shadowing. The voids between the columns lead to instabilities in the properties of the film when it is exposed to a humid atmosphere since they will fill with water by capillary action. Figure 6.2(a) shows the effect of such water adsorption on the spectral transmittance of a layer of ZrO_2 before and after venting to a humid atmosphere. The shift in the curve to higher wavelengths is interpreted as an increase in the film refractive index due to the void refractive index increasing from that of vacuum (1.0) to that of water (1.33). Under IAD conditions, the adatoms on the growing film surface have greater mobility, the columnar microstructure is disrupted, and stable dense films are produced. Figure 6.2(b) shows that an IAD ZrO_2 film does not adsorb water on venting the film to humid air. This densification effect has been verified many times and has been explained theoretically in Reference [3].

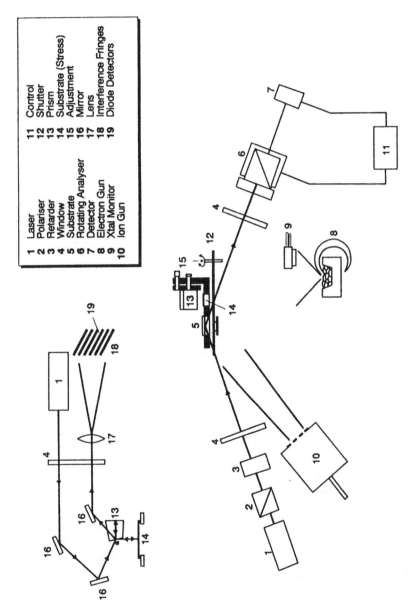

Figure 6.1 Schematic diagram of the deposition equipment and *in situ* ellipsometer and Ennos type stress interferometer.

1	Laser	11	Control
2	Polariser	12	Shutter
3	Retarder	13	Prism
4	Window	14	Substrate (Stress)
5	Substrate	15	Adjustment
6	Rotating Analyser	16	Mirror
7	Detector	17	Lens
8	Electron Gun	18	Interference Fringes
9	Xtal Monitor	19	Diode Detectors
10	Ion Gun		

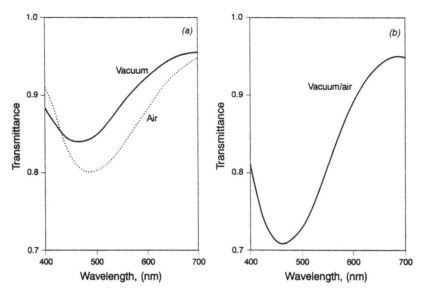

Figure 6.2 Spectral tramsmittance of ZrO₂ measured in vacuum immediately after deposition (solid line) and on venting to a humid atmosphere (dotted line): (a) evaporation only, (b) IAD.

Optimization of the deposition conditions can be tedious if *ex situ* characterization is required since the deposition rate, ion species, ion energy, and ion current density can all be varied independently. For optical films the optimum conditions usually correspond to those which give the highest refractive index. Real-time ellipsometric monitoring can provide data on the index of the near surface region. Figure 6.3a shows the raw data obtained during the deposition of a Ta_2O_5 film under different bombardment conditions, and Figure 6.3b shows the refractive index versus ion current density curve deduced from a limited number of such experiments. In many cases, the optical properties of the film are strongly related to other properties of the layer so the technique gives a method of rapidly mapping out the deposition parameter space.

The stress in a thin film coating can be a major contributing factor to other mechanical properties such as hardness and adhesion. However, the stress in a film is known to be dependent on many of the deposition conditions and particularly the ion-bombardment parameters. Ideally, the stress should be measured *in situ* during film growth to avoid any changes that may occur upon exposure of the deposited film to the atmosphere. The combination of stress interferometry and ellipsometry is a powerful method of monitoring and controlling the properties of optical materials deposited by IAD [6]. Figure 6.4a shows the raw ellipsometric data obtained during the deposition of a SiO_x film under different IAD conditions.

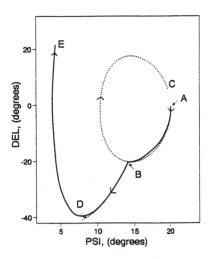

Figure 6.3a Plot of the ellipsometric angles DEL and PSI for thin film deposition of Ta₂O₅. Deposition started at position A and the conditions were changed near positions B and D. The dashed line is a theoretical plot assuming homogeneous layers between A to B, and B to D.

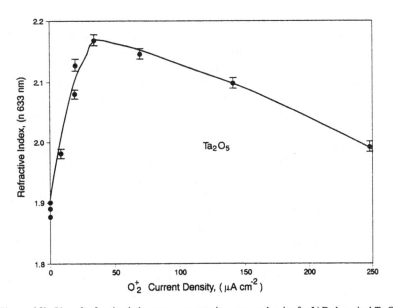

Figure 6.3b Plot of refractive index versus oxygen ion current density for IAD deposited Ta₂O₅ films.

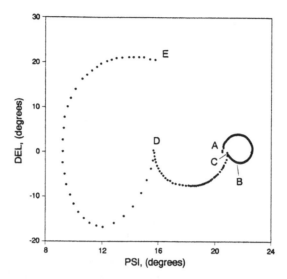

Figure 6.4a Plot of ellipsometric angles DEL and PSI for the growth of SiO$_x$ deposited: (a) plot A–B, under 100 eV IAD, heavy bombardment, refractive index, $n = 1.46$, (b) plot B–C, 100 eV IAD, reduced bombardment, $n = 1.46$, (c) plot C–D, reactive evaporation in O$_2$, $n = 1.62$, (d) plot D–E, evaporation of SiO at base vacuum, $n = 1.905$.

Figure 6.4b is a plot of the total film stress as a function of the film thickness measured ellipsometrically. Under different conditions, the film is growing with varying degrees of stress ranging from compressive to tensile, although in this example the total film stress is always compressive.

Ion assistance is also known to improve the adhesion of films to substrates and other films. Notably, *in situ* monitoring has been used to investigate the nucleation, early growth, and coalescence of gold films. It was shown that the coalescence thickness was reduced from about 18 nm for an evaporated film to 4 nm for an oxygen ion-assisted film, and the IAD film completely "wet" the substrate whereas the substrate-film interface was voided in the evaporation case. *In situ* ion scattering spectroscopy (ISS) is a powerful complementary technique since it provides information as to the elemental composition of the outermost monolayer. Figure 6.5 shows an ISS spectra obtained during the deposition of a gold film. The presence of the Si and O peaks confirms that the gold has not yet completely covered the substrate surface. The combination of these *in situ* techniques has shed much light on the nucleation, growth, densification, and compound synthesis mechanisms of film growth by IAD [3].

Advanced *in situ* monitoring will play an increasingly important role in industrial physical vapour deposition (PVD) systems. Figure 6.6 shows an example of ellipsometric monitoring of a two layer anti-reflection inter-

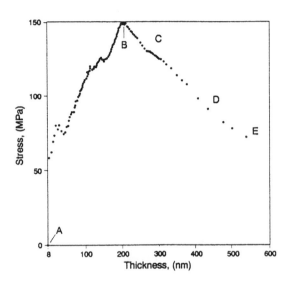

Figure 6.4b The corresponding compressive stress versus thickness plot for the whole film.

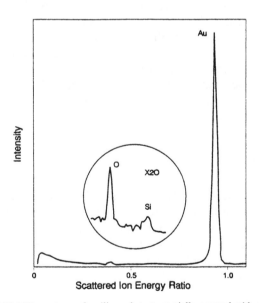

Figure 6.5 ISS spectrum of a silica substrate partially covered with an Au film.

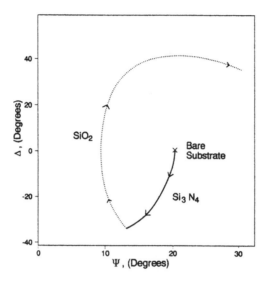

Figure 6.6 Ellipsometric monitoring of a two-layer anti-reflection coating.

ference coating. The data can deduce the refractive index and thickness of each layer in real time to high accuracy. If any deviation from the theoretical design is detected during deposition, the multilayer can be reoptimized for the following layers. Monitoring of the compound synthesis of materials such as TiN, whose properties are very dependent on operating conditions, can inform the operator in real time if deviation from the ideal is occurring. Hartley and Folkhard [9] have recently shown how differential ellipsometry can be used to monitor and control the deposition of cadmium mercury telluride.

6.4 CONCLUSIONS

Ion-assisted deposition is one of the many ion-based PVD techniques, which have made substantial improvements in many of the physical properties of deposited films. *In situ* real-time monitoring can be used both as a means of mapping out the parameter space rapidly and as a production process control. It is a simple operation to make such processes operate in a closed-loop manner.

6.5 REFERENCES

1 Hirsch, E. H. and Varga, I. K. "The Effect of Ion Irradiation on the Adherence of Germanium Films," *Thin Solid Films*, 52:445–452, 1978.

2 Martin, P. J. "Ion-Based Methods for Optical Thin Film Deposition," *J. Mater. Sci.*, 21:1–25, 1986.

3 *Handbook of Ion Beam Processing Technology: Principles, Deposition, Film Modification and Synthesis*, J. J. Cuomo, S. M. Rossnagel, and H. R. Kaufman, eds., Noyes Publications, New Jersey, 1989.

4 Netterfield, R. P., Martin, P. J. and Muller, K.-H. "*In situ* Optical Monitoring of Thin Film Deposition," *In-Process Optical Measurements*, SPIE, 1012:10–15, 1988.

5 Netterfield, R. P., Martin, P. J., Sainty, W. G., Duffy, R. M. and Pacey, C. G. "Characterization of Growing Thin Films by *in situ* Ellipsometry, Spectral Reflectance and Transmittance Measurements, and Ion-Scattering Spectroscopy," *Rev. Sci. Instrum.*, 56:1995–2003, 1985.

6 Martin, P. J. and Netterfield, R. P. "*In situ* Stress Measurements of Ion Assisted MgF_2 and SiO_x Thin Films," *Appl. Phys. Lett.*, submitted, 1990.

7 Kaufman, H. R. and Robinson, R. S., in Reference [3], pp. 8–20 and pp. 39–57.

8 Ennos, A. E. *Appl. Opt.*, 5:51–61, 1966.

9 Hartley, R. and Folkhard, M. A., "Differential Ellipsometry for Monitoring the Growth of Thin Films," *Conf. Proc. 8th AIP Congress*, p. 261, 1988.

CHAPTER 7

Refractory Metal and Alloy Coatings Deposited by Magnetron Sputtering

D. M. TURLEY[1]

7.1 INTRODUCTION

REFRACTORY metal and alloy coatings have applications at elevated temperatures where improvements in wear corrosion, oxidation, or erosion resistance are required. Chromium can be electrodeposited from aqueous electrolyes; however, to electrodeposit other refractory metals, molten salt electrolytes are normally required. The FLINAK process [1] is a molten salt process and is used to deposit refractory metals such as niobium and tantalum. The process, however, is operated at approximately 800°C, which for hardened and tempered steel substrates, is well above the tempering temperature, and this results in a degradation in the steel mechanical properties.

Physical vapour deposition (PVD) processes such as ion plating and magnetron sputtering are methods whereby refractory metals and alloys can be deposited at temperatures below steel tempering temperatures [2–4]. Movchan and Demchisin [5] have reported from their studies of thick evaporated coatings that the nature of the microstuctures produced is a function of T/T_m, where T is the substrate temperature during deposition and T_m the melting point of the deposited metal. They identified three distinct microstructures, which they designated as: Zone 1 ($T/T_m > 0.3$), which consisted of tapered crystals; Zone 2 ($0.3 < T/T_m < 0.45$), which had a columnar structure; and Zone 3 ($T/T_m > 0.5$), which had an equiaxed grain structure. Thornton [6] subsequently extended this model to include the effect of gas pressure and subsequently identified a transition

[1]Materials Division, Materials Research Laboratory, MRL-DSTO, Melbourne, P.O. Box 50, Ascot Vale, Victoria 3032, Australia.

zone (Zone T) between Zone 1 and 2 structures consisting of densely packed fibrous grains. The gas pressure affected the T/T_m ratio at which these zones formed, with lower gas pressure resulting in the zones forming at lower T/T_m ratios.

Columnar type structures are considered to be produced by geometric shadowing of the incident atoms by atoms already deposited on the substrate [7]. By bombarding the surface with ions by applying a negative bias [8,9] or by bombarding it with atoms with a fast atom beam (FAB) source [4], surface adatom mobility is enhanced, resulting in the rearrangement of the coating atoms, smaller shadowing affects, and the reduction or elimination of columnar type structures. Brooker [10] and Mawella and Sheward [11] have reported that, by simultaneously depositing two differently sized metal atoms, the self-shadowing process is disturbed and the migration of atoms enhanced, resulting in less void formation and a denser microstructure.

Dirks and Leamy [7], however, have concluded that shadowing effects will be enhanced and columnar grain structures separated by voids more pronounced as the angle of incidence of the coating flux relative to the substrate normal increases. Moreover, they also reported [7] that the columnar grains are oriented away from the substrate normal towards the direction of the coating flux. This angle is less than the angle of incidence of the coating flux but increases as the angle of incidence increases. Thus, when coating an irregularly shaped surface with a unidirectional coating flux, a range of coating microstructures would be produced.

In the present study, the effect of the angle of coating flux incidence on coating microstructure was studied by coating steel substrates oriented normal and parallel to the dominant coating flux produced from magnetron sputtering targets of refractory metals and alloys. The effects of subjecting the coating during deposition to atom bombardment from a FAB source, negative bias and alternatively depositing metals, which had limited solubility in each other and were mutually soluble and had dissimilar and similar atomic radii respectively, on coating microstructure was also investigated.

7.2 EXPERIMENTAL PROCEDURE

The substrates were of polished gun steel and were in the form of flat plates (40 mm × 40 mm) and 30 mm inner diameter rifled tubes.

Prior to coating by magnetron sputtering, the substrates were sputter cleaned with the FAB source using Ar atoms. For flat substrates, the sputter cleaning time was 2 hr as it had been previously determined [12] that this removes oxide layers from the substrate resulting in good adhesion of the coating. For rifled tubular substrates where the line of sight path of the atoms from the FAB source is tangential to the substrate, a cleaning time of 3–4 hr was used. The substrates could be rotated about the central axis so that, after FAB cleaning, they could be quickly positioned over a magnetron. Alternatively, the substrates could be oscillated between the magnetrons to produce alloy coatings or between the magnetrons and the FAB source to produce alternate coating and FAB source bombardment. A dc bias could also be applied to the substrates. A schematic sketch showing the position of the substrates with respect to the magnetrons is shown in Figure 7.1. Flat substrates and rifled tubes were positioned 80 and 38 mm above the magnetrons, respectively. Chromium, niobium, tantalum, molybdenum, tungsten, and austenitic stainless steel (SS) targets were used. Targets of SS were used instead of iron because, being nonmagnetic, they could be magnetron sputtered.

FAB cleaning was carried out with an Ion Tech FAB source at 5–15 mA and a pressure of $\sim 10^{-3}$ mbar. The magnetron sputtering conditions were: voltage 400–600 V, current 0.2–0.65 A and chamber pressure 3–8 × 10^{-3} mbar. Simultaneous operation of the FAB source, using Ar atoms, and the magnetrons was carried out at pressures of 3–8 × 10^{-3} mbar.

The surface topography and fracture cross sections of coatings were examined in a scanning electron microscope (SEM). Analyses of the

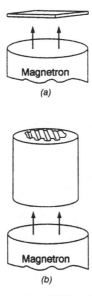

Figure 7.1 Schematic sketch showing position of (a) flat plates and (b) rifled tubes, with respect to the magnetrons.

coatings were carried out on metallographically polished cross sections in a Cameca Electron Probe Microanalyser.

Alloy coatings representative of the types deposited were subjected to X-ray diffraction analysis to determine the phases present. A Nonius Weissenberg camera was used and the coated surface illuminated with a collimated beam of Co K-alpha radiation. The specimen was oscillated about the centre of the camera through 140 degrees, and a set of diffraction rings was obtained over the entire cylindrical recording film.

7.3 RESULTS AND DISCUSSION

7.3.1 COATING OF FLAT PLATES

Visually, these coatings had a bright, specularly reflecting surface; however, examination in the SEM showed that coating defects were present, the most prevalent type being nodular growths [10,13]. These growth defects frequently nucleate at the substrate surface and appear in the coating as inverted cones with their apex at the substrate and their domed bases forming the tops of the nodules protruding above the coating surface. These growth defects are not firmly adherent to the remainder of the coating, and they are easily detached leaving holes. The surface of the coatings between the defects varied from being smooth and relatively featureless to slightly faceted (Figure 7.2). SEM examination of fracture sections revealed that a fine columnar type grain structure was present (Figure 7.3).

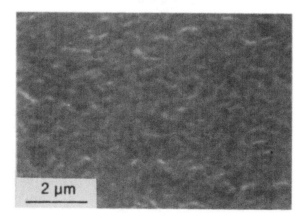

Figure 7.2 Scanning electron micrograph of the surface of a noibium coating. Surface is slightly faceted. Flat plate.

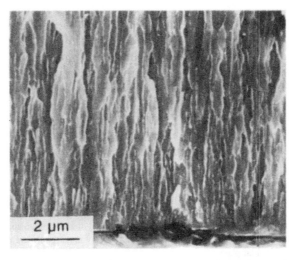

Figure 7.3 Scanning electron micrograph of a fracture section of a niobium coating. A columnar type grain structure is present. Flat plate.

Chromium coatings produced by oscillating the substrate between the magnetron and the FAB source or by coating at high deposition rates (10 μm hr^{-1}), were virtually free from nodular growths. Visually, these coatings had a matte-etched appearance, and examination in the SEM showed that the surface was faceted. Moreover, examination of fracture sections showed that columnar structures were no longer present (Figure 7.4, cf. Figure 7.3). A reduction in the incidence of nodules together with grain refinement of the coating has also been reported [8,10] to occur when a negative bias is applied to the substrate during sputter coating (sputter ion plating). The refinement of the grain structure and absence of nodules at high deposition rates is attributed to the increased energy of the sputtered atoms striking the substrate, which increases adatom surface mobility, enabling rearrangement of the coating atoms to occur, reducing shadowing effects. Bombardment with the FAB source removes material from the substrate, and it is considered that many nodule initiation sites are removed. Argon atom bombardment also enhances surface adatom mobility of the coating atoms, reducing shadowing effects resulting in the refinement of the columnar type structures.

From Thornton's [6] structural zone model the coatings deposited at the lower deposition rate (2 μm hr^{-1}) and 4 $\times$ 10^{-3} mbar without FAB bombardment should have a Zone T structure ($T/T_m \sim 0.1$–0.2) consisting of tightly packed fibrous grains, and this corresponds to the fine columnar grain structures observed (Figure 7.3). It is not possible to predict the effects of atom bombardment on coating morphology from Thornton's

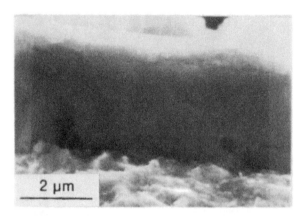

Figure 7.4 Scanning electron micrograph of a fracture section of a chromium coating produced by alternate deposition and argon atom bombardment. Note, columnar grains are no longer present (cf., Figure 7.3). Flat plate.

model; however, at high T/T_m ratios where surface adatom mobility is high, the model predicts equiaxed grain structures (Zone 3).

7.3.2 COATING OF THE BORES OF 30 MM RIFLED TUBES

To do this, the coating flux was directed up the tube so that the angle of incidence of the coating flux was approximately parallel to the internal surfaces being coated (Figure 7.1). This resulted in significantly less deposition than when the coating flux was normal to the substrate. Coatings had a very open columnar type structure, and distinct columnar crystals could be observed, which were larger at the end of the tube closest to the magnetron (bottom end). Therefore, for comparative purposes, all scanning electron micrographs were taken at the bottom end on the land tops. A niobium coating is shown in Figure 7.5, and distinct columnar crystals are present separated by large voidal regions. The tops of the crystals were faceted, and examination of the sides indicated a dendritic-like growth pattern. This type of structure would correspond to a Zone 1 structure according to Thornton's structural zone model, and by comparing Figure 7.5 and Figure 7.2, the effect of the angle of incidence of the coating flux on coating morphology is clearly apparent.

A transverse fracture section of the coating is shown in Figure 7.6 and should be compared with the fracture section of the coating produced at normal flux incidence (Figure 7.3). Individual columnar crystals separated by voids are clearly apparent (Figure 7.6) moreover, fracture has occurred via the voids not through the columnar crystals. These results are in agreement with Dirks and Leamy (Figure 7.7) who reported that, as the angle

Figure 7.5 Scanning electron micrograph of the surface of a niobium coating. Distinct columnar crystals are present. Rifled tube.

Figure 7.6 Scanning electron micrograph of a transverse fracture section of the niobium coating shown in Figure 7.51. Columnar crystals are present.

Figure 7.7 Scanning electron micrograph of the surface of a Nb-Cr coating produced by alternate deposition and argon atom bombardment. The coating is dense. Rifled tube.

of incidence of the coating flux relative to the substrate normal increases, more pronounced and distinct columnar grain structures separated by voids are produced. Alternate coating and argon atom bombardment with the FAB source had the effect of densifying the structure shown in Figure 7.4; however, a columnar crystal structure could still be discerned.

In order to further densify the coating, FAB bombardment was combined with the alternate deposition of two different metals (alloying) using two magnetrons. The alloys deposited are given in Table 7.1. Two types of alloys were studied. In the first, the metals had only limited solubility in each other under equilibrium conditions [14] and different atomic radii [15]. Since the metals had only limited solubility in each other, it was considered that atoms deposited from the two magnetrons would be less likely

TABLE 7.1. **Characteristics of Alloys Deposited.**

Alloy	Composition (wt %)	Mutually Soluble (Equilibrium Conditions)	Structure	Ratio of Atomic Radii	Atomic Radii (Å)
Nb-Cr	Nb (54), Cr (47)	No	Dense	Nb/Cr (1.15)	Mo (1.40)
Nb-SS	Nb (54)	No	Dense	Nb/Fe (1.15)	W (1.41)
W-SS	W (77)	No	Dense	W/Fe (1.10)	Ta (1.47)
Nb-Ta	Nb (27), Ta (73)*	Yes	Open	Nb/Ta (1.00)	Cr (1.28)
Nb-W	Nb (32), W (68)	Yes	Open	Nb/W (1.04)	Nb (1.47)
Nb-Mo	Nb (37), Mo (63)	Yes	Open	Nb/Mo (1.05)	Fe (1.28)

*Approximate.

to build up continuously to form columnar crystal type growth structures [4]. Moreover, it has been reported [10,11] that simultaneously depositing two metals that have different atomic radii reduces self shadowing effects, producing denser microstructures. Alloys that meet these requirements and were investigated in the present study are Nb-Cr, SS-Nb and SS-W. Conversely, it was considered that, when two metals are mutually soluble [14], there would be a tendency for the buildup of deposited atoms to be continuous, similar to when a single metal is deposited, and thus the open type columnar crystal structures would still be formed. Alloys that meet this requirement and were investigated were Nb-Ta, Nb-W, and Mo-Nb. Moreover, these metals have similar atomic radii (Table 7.1), and therefore shadowing should not be affected.

The results are summarized in Table 7.1. Alloys formed from metals that had only limited solubility in each other and different atomic radii produced dense coatings in which the columnar structures were not as pronounced. For alloys formed from metals that were soluble in each other under equilibrium conditions and that had similar atomic radii, the coarse columnar crystal structure persisted. A scanning electron micrograph of the surface of the Nb-Cr coating is shown in Figure 7.7. The open columnar crystal structure is no longer present (cf., Figure 7.5), and the structure is quite dense. Surface features are present and are considered to be the tops of closely packed columnar grains. A fracture section of the coating showed that a densely packed columnar type structure was present (Figure 7.8, cf., Figure 7.6). The surface of the Nb-W coating is shown in Figure 7.9, and in this case an open columnar crystal growth structure sep-

Figure 7.8 Scanning electron micrograph of a transverse fracture section of the Nb-Cr coating (Figure 7.7). A dense columnar structure is present.

Figure 7.9 Scanning electron micrograph of Nb-W coating produced by alternate deposition and argon atom bombardment. Rifled tube.

arated by large voids is still present. The fracture sections of these types of coatings showed columnar crystals separated by large voids. Applying a dc bias, as well as bombarding with the FAB source, tended to densify the structures produced for both types of alloy.

Longitudinal fracture sections of the Nb, Nb-Cr, and Nb-Ta alloys are shown in Figure 7.10(a)–(c), respectively. As reported by Dirks and Leamy

(a)

Figure 7.10 Scanning electron micrographs of longitudinal fracture sections through coatings of (a) niobium produced by alternate deposition and argon atom bombardment. The columnar structures are oriented towards the direction of the incident coating flux (right-hand side). Rifled tubes.

(b)

(c)

Figure 7.10 (continued) Scanning electron micrographs of longitudinal fracture sections through coatings of (b) Nb-Cr and (c) Nb-Ta produced by alternate deposition and argon atom bombardment. The columnar structures are oriented towards the direction of the incident coating flux (right-hand side). Rifled tubes.

[7], the grain structure is oriented towards the direction of the coating flux. For the Nb coating, distinct columnar crystals can be clearly seen [Figure 7.10(a)]. For the Nb-Cr alloy, there is a closely packed columnar grain structure [Figure 7.10(b)], whereas the structure of the Nb-Ta alloy is intermediate between the two [Figure 7.10(c)]. The angles at which the columnar structures were oriented away from the substrate normal towards

the direction of the coating flux were measured and found to be approximately similar (50–55°) for the Nb, Nb-Cr, and Nb-Ta coatings. Thus,
alloying does not appear to have significantly altered the angle at which the
columnar structures are oriented, and the orientation was not affected by
alloy type. That is, whether the metals were soluble in each other under
equilibruim conditions or had differences in atomic radii did not affect the
orientation of the columnar grain structure.

For the Nb-Ta alloy, which forms a continuous solid solution under equilibrium conditions [14], X-ray diffraction resulted in reflections whose d
values corresponded to those of Nb and Ta; their unit cells are almost identical in lattice parameter, and therfore their diffraction lines almost
superimpose. The broad nature of the lines in the diffraction pattern precludes the determination of whether there are both Nb or Ta structures, or
a single Nb-Ta bcc structure present. For the Nb-Cr, SS-W, and SS-Nb
alloys that are not mutually soluble in each other, intermediate phases are
formed under equilibrium conditions according to the phase diagrams
[14]; however, none of these phases were identified by X-ray diffraction.
This indicates that nonequilibrium alloy structures can be produced by
magnetron sputtering, and this effect has also been observed in PVD
coatings by other investigation [16].

7.4 CONCLUSIONS

(1) Metal coatings deposited at normal incidence (flat plates) at low to
    moderate deposition rates had smooth to slightly faceted surfaces and
    a fine columnar grain structure. Nodular growth defects were also
    present.

(2) Metal coatings deposited at normal incidence at higher deposition
    rates or by alternate coating and argon atom beam bombardment had
    increased surface faceting, no columnar grain growth, and virtually
    no nodular growth defects.

(3) Metal coatings deposited at parallel incidence (rifled tubes) at low to
    moderate deposition rates had an open structure consisting of distinct
    columnar crystals.

(4) Alloy coatings deposited at parallel incidence by alternate deposition
    of two metals, which had limited solubility in each other and different
    atomic radii, together with argon atom beam bombardment, produced
    dense coherent coatings. A columnar grain structure was still discern-
    ible.

(5) Conversely, alloy coatings deposited at parallel incidence by alternate
    deposition of two metals, which were mutually soluble and had simi-

lar atomic radii, together with argon atom beam bombardment, produced coatings that had an open structure and distinct columnar crystals.

(6) When coating at parallel incident, the columnar structures were oriented towards the direction of the incident coating flux. Alloy type did not significantly affect the angle of orientation.

7.5 ACKNOWLEDGEMENTS

The author would like to thank Mr. C. Townsend for his asssistance in producing the coatings and preparing the fracture sections. The assistance of Messrs. J. Russell, A. Gunner, and I. McDermott who, respectively, carried out the scanning electron microscopy, EPMA analysis, and X-ray diffraction is also gratefully acknowledged.

7.6 REFERENCES

1 Ahmad, I., Janz, G. J., Spiak, J. and Chen, E. S. "Electrodeposition of Refractory Metal Coatings (Tantalum and Columbium) from Fused Salt Electrolytes," *Proc. Tri-Service Gun Tube Wear and Erosion Symp.*, Dover, New Jersey: 25–27 October, pp. iv, 299–313, 1982.

2 Glue, D. R., Sheward, J. A., Young, W. J. and Gibson, I. P. "Erosion Resistant Coatings for Gun Bore Surfaces," RARDE Report 1/81, June. 1981.

3 Gibson, I. P. "Erosion Resistant Coatings of Tubes by Physical Vapour Deposition," *Thin Solid Films*, 83:27–35, 1981.

4 Turley, D. M. "Development of Erosion Resistant Refractory Metal and Alloy Coatings by Magnetron Sputtering," *Surface and Coatings Technology*, 39/40: 135–142, 1989.

5 Movchan, B. A. and Demchisin, A. V. "A Study of the Structure and Properties of Thick Vacuum Condensates of Nickel, Titanium, Tungsten, Aluminium Oxide and Zirconium Dioxide," *Fry. Met. Metalloved*, 28:653–660, 1969.

6 Thornton, J. A. "Influence of Apparatus Geometry and Deposition Conditions on the Structure and Topography of Thick Sputtered Coatings," *J. Vac. Sci. Technol.*, 11:666–670, 1974.

7 Dirks, A. G. and Leamy, H. J. "Columnar Microstructure in Vapor-Deposited Thin Films," *Thin Solid Films*, 47:219–233, 1977.

8 Mattox, D. M. and Kominiak, G. J. "Structure Modification by Ion Bombardment during Deposition," *J. Vac. Sci. Technol.*, 9:528–532, 1972.

9 Bland, R. D., Kominiak, G. J. and Mattox, D. M. "Effect of Ion Bombardment during Deposition on Thick Metal and Ceramic Deposits," *J. Vac. Sci. Technol.*, 11:671–673, 1974.

10 Brooker, C. J. "The Deposition of Thick Refractory Metal Coatings on Gun Steel Substrates by Sputter Ion Plating," RARDE Memorandum 92/84, 1984.

11 Mawella, K. J. A. and Sheward, J. A. "The Effect of Co-Deposition of Metals on the Microstructure and Properties of Sputtered Alloy Coatings," *Int. Conf. on Ion and Plasma Assisted Techniques (IPAT)*, Geneva, pp. 70–75, 1989.

12 Turley, D. M. "Development of Erosion Resistant Coatings by Magnetron Sputtering," *Proc. Int. Tribology Conf.*, Melbourne, Victoria, 2–4 December, pp. 80–86, 1987.

13 Spalvins, T. and Brainard, W. A., "Nodular Growth in Thick-Sputtered Metallic Coatings," *J. Vac. Sci. Technol.*, 11:1186–1192, 1974.

14 *Metals Handbook, Vol. 8, 8th Edition,* American Society of Metals, Metals Park, Ohio, pp. 251–376, 1973.

15 *Smithells Metals Reference Book (Sixth Edition).* Table 4.21, E. A. Brands, ed., Butterworths, 1983.

16 Thornton, J. A. "High Rate Thick Film Growth," *Annual Rev. of Mat. Sci., Vol. 7,* R. A. Huggins, ed., Annual Reviews Inc., California, pp. 239–260, 1977.

CHAPTER 8

A Study of Coatings Produced by Sharc PVD Arc Source

P. JEWSBURY[1]
P. LAMBRINEAS[1]
K. JACKSON[1]

8.1 INTRODUCTION

PHYSICAL vapour deposition (PVD) sources are important commercially because they can be used to deposit [1] adherent, well-consolidated thin protective films in a well-controlled manner at "low" substrate temperatures ($<400°C$). Depending on which type of system is employed, the main drawbacks may arise from the high capital cost, fixturing, and load limitations associated with exposure requirements to ancillary energy beams and "long" cleaning and deposition times.

In this chapter, studies of coatings produced using a unique anodic-cathodic arc source are reported. Arc sources are the most productive of all PVD sources, with the highest degree of ionization of the emitted flux, thereby giving them a possible commercial edge (ancillary energy beams can be avoided by using suitable biasing allowing greater packing of the chamber and fast processing times). The production rate from an arc source is sufficiently high that it can be measured in practical units of gas flow, standard cubic centimetres per minute (SCCM). Thus, for a titanium random arc of I amperes, operating in p millitorr of nitrogen, the production rate is approximately (to within 10 SCCM) given by [2]

$$\text{Flux} \approx \exp(-0.124 - 0.0167p)I \qquad \text{(SCCM)} \qquad (8.1)$$

Conventional arc sources are cathodic random or steered arcs. Steered arcs use electromagnetic fields to constrain the trajectory of the arc on the

[1]Materials Research Laboratory, (MRL)—DSTO, P.O. Box 50, Melbourne 3032, Victoria, Australia.

cathode surface and can be transformed into fully controllable alloy sources [3]. This chapter is principally concerned with the properties of thin coatings of TiN produced by a new random arc system. These have been studied by several techniques including X-ray diffraction and optical photometry.

8.2 EQUIPMENT

An unusual arc source dubbed the Sharc has been developed and tested. This consists of an anodic rod (containing a gas feed) and cathodic plate as shown in Figure 8.1. Full details of the apparatus have been described elsewhere [4]. Several studies have been completed of the details of the arc process and nature of the emitted flux. These include a spectrographic analysis of the light emitted from the arc region providing estimates of the temperature of the bound electrons in the fluxes and their degree of ionization [5], studies of the ion fluxes and distribution using a Faraday cup, and studies of macroparticle production by examining film roughness [4]. Thus, it has been found that this design provides excellent control over the macroparticle content and ionicity of the emitted fluxes from both anode and cathode. Its main drawback over existing arc designs relates to the obtrusive and obstructive presence of the anode, which carries gas and current feeds into the chamber.

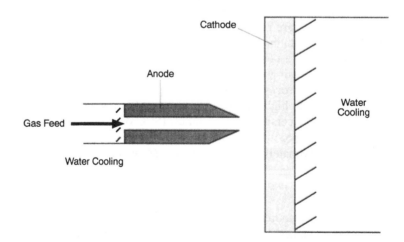

Figure 8.1 Schematic diagram of active components of the Sharc arc. The hollow anode introduces gases directly into the arc region. During the operation of the arc, the anode becomes hot and, by radiation, introduces a temperature gradient on the cathode surface. Both anode and cathode evolve titanium atoms and ions.

8.3 COATING TECHNIQUE

Coatings were prepared in a large vacuum system fitted with the Sharc arc in the upper part of the chamber. The chamber was equipped with rotary and diffusion pumps. High purity nitrogen and argon gases were used. The specimens were located in a set of three enclosed boxes. An exterior handle enabled the boxes to be sequentially exposed either to Argon cleaning or the fluxes from the arc at a position 220 mm below the anode rod and 280 mm from the cathode centre. A thermocouple was embedded in the base plate on which the specimen boxes rested. This gave an indication of the specimen temperature.

Generally, each of the boxes contained a specimen, and the procedure was as follows. The specimens were sequentially Argon cleaned for 1/2 hr at 15 millitorr pressure in a 600-volt glow discharge. With the specimens enclosed in the boxes, the arc was struck and allowed to stabilise for 1 hr. The boxes were then exposed sequentially for 1 hr each, with the arc conditions being changed slightly between each exposure. Finally, the specimens were allowed to cool under vacuum before being removed from the chamber.

8.4 STRUCTURE AND COMPOSITION

Many of the specimens were examined by X-ray diffraction. The only phase of TiN observed was the fcc δ phase. XRD was used to determine the lattice constant normal to the film surface and, in future work, is being used to measure the residual stresses in the films using the $\sin^2 \phi$ technique. A considerable variation in lattice constant with process conditions was observed. For stress free $Ti_{1-x}N_x$ ($0.28 < x < 0.5$) prepared at high temperatures, the lattice constant, a, varies with concentration as [6]

$$a = 4.160 + 0.16x \tag{8.2}$$

At room temperature, however, the δ-phase only extends down to $x = 0.33$ where the tetragonal ϵ subnitride phase is the stable phase. In an otherwise defect-free lattice, the incidence of voids and the changes in lattice spacing will produce a change in the electronic bandstructure and, hence, a change in the majority of physical properties. Many properties could therefore be a good indication of the composition of a $Ti_{1-x}N_x$ coating. Amongst the simplest is optical colourimetry, which is described more fully in the next section. This can be used to ascribe quantifiable colour coordinates to a thin film.

The XRD lattice parameter is plotted against the y' % colour coordinate in Figure 8.2. The colour coordinate and lattice parameter are clearly

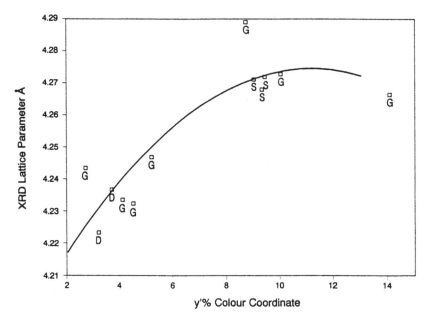

Figure 8.2 Correlation between the lattice parameter in angstroms determined by X-ray diffraction and the y' % colour coordinate determined by a Varian spectrophotometer. Observed colour is indicated by the letters D for dark, G for golden or brassy, and S for silvery.

strongly correlated. Coatings are visibly seen to range from silvery (excess incident Ti, $x < 0.5$) to dark brown (excess incident nitrogen, $x > 0.5$), but this ordering is not perfectly duplicated by the y' % colour coordinate. To illustrate this, the data points in Figure 8.1 are labelled D (for dark brown/black), G (for golden or brassy), and S (for silvery). Note that a single colour coordinate does not uniquely differentiate between these colour groupings. Furthermore, according to Equation (8.1), the dark excess nitrogen coatings should have the larger lattice constants and the silvery coatings the smaller ones. This is contrary to the results of Figure 8.2. Thus, for these arc coatings, the lattice constant and physical parameters (correlated by the deposition process), such as stress and microstructure, have the dominant effect in determining the y' % colour coordinate. The variation of lattice constant with composition contrary to Equation (8.2) is thus a reflection of the deposition conditions, producing a nonequilibrium structure.

8.5 SPECIMEN COLOUR

Apparent colour is generally a good guide to coating conditions. As discussed earlier, if coatings are grown when the Ti flux is low, they are dark (strong absorption across the visible band). As the Ti flux is increased, the absorption falls and the coatings become, firstly, brassy coloured (greeny blue absorption), then golden-yellow (predominantly blue absorption), and finally silver (less absorption, which is strongest in the blue region of the visible spectrum).

To quantify this progression, measurements have been made of the absorbance and colour coordinates of several TiN films using a Varian spectrophotometer. All results reported here involve the colour of the coatings under D65 simulated daylight. The colour coordinates [7] mimic the subjective colour observed by an eye possessing three colour receptors. They are normalized to unity so as to reflect the hue and tone of the colour, not its brightness. Following standard procedures, these colour coordinates are transformed into a set (x,y,z) of chromaticity coordinates, which are always positive. Thus, each coordinate varies across the spectrum with the x and y coordinates having the largest spread. However, a single coordinate cannot uniquely define a colour.

The reflection of light across the visible spectrum for three coatings produced by an 115 ampere arc running in 5.5, 10.6, and 16.7 millitorr pressure of nitrogen is shown in Figure 8.3a. Maximum absorption of light occurs at low wavelengths. As the pressure of nitrogen increases (eventually leading to a nitrogen-rich coating), the wavelength at maximum absorption increases (as depicted in Figure 8.3b) and the absorption band broadens

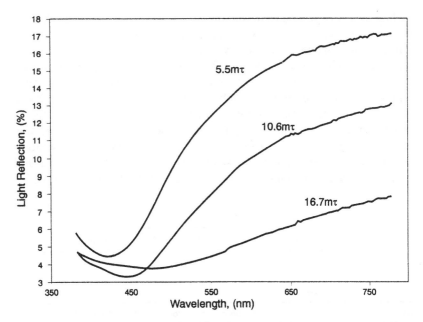

Figure 8.3a Reflection of D65 simulated daylight from thin TiN films produced by a 115 ampere Sharc arc source operating in pure nitrogen at pressures of 5.5, 10.6, and 16.7 millitorr.

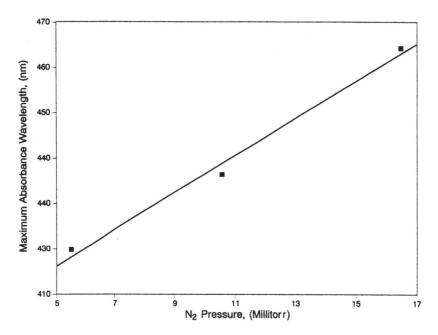

Figure 8.3b Shift of maximum in absorbance (minimum in reflection) with nitrogen pressure for the films shown in Figure 8.3a.

leading to a less pronounced colour. For titanium-rich coatings, the absorption peak has shifted to wavelengths in the ultraviolet.

It was found that the colour coordinates were very sensitive to contaminant gases, surface cleaning, and surface finish. The consequences of poor surface cleaning are illustrated in Table 8.1. Three samples were polished, degreased, and coated in the chamber with a 135 ampere arc and 80 volt bias. All specimens were argon cleaned for 1 hr and then held under vacuum for 24 hr prior to the deposition. Immediately before the arc was started, specimen c2 was cleaned again for fifteen minutes. The coating procedure on the specimens was sequential starting with c1 then c2 and finally c3. The coating on samples c1 and c3 appeared to be identical, in-

TABLE 8.1.

Specimen ID	$y\%$ Colour	Second Clean
c1	9.6	no
c2	12.5	yes
c3	9.2	no

Specimens c1 and c3 argon cleaned but left under vacuum for 24 hr. They exhibit a different colour to specimen c2 which was also cleaned again prior to the deposition.

dicating that deposition order was not relevant. c2 differed significantly from the other two films.

The effects of oxygen and hydrogen contamination of the films is shown in Figure 8.4. Oxygen will substitutionally replace nitrogen. The presence of oxygen should be expected to reduce the lattice constant ($a = 4.17$ Å for TiO compared to $a = 4.23$ Å for TiN). However, for the samples shown in Figure 8.4, both the colour coordinate and lattice parameter were found to increase. Hydrogen is also incorporated in the growing films significantly changing the colour coordinate. These changes are, however, much harder to detect by eye. Thus, a colour analysis could provide a useful indication of contamination of a growing TiN film.

However, colour coordinates are even more sensitive to substrate surface finish. The $y\%$ colour coordinate of TiN films prepared in 15 millitorr of nitrogen at arc currents from 115 to 165 amperes is shown in Figure 8.5. Note that the colour coordinate of films on rough surfaces is very different from that on a polished surface.

Incident ionic fluxes are not uniform on rough surfaces, which also impede the surface diffusion of adsorbed atoms. Thus, on rough surfaces, there are many growth defects, and the growth pattern may be columnal with many voids. A high incidence of defects with a size comparable to the

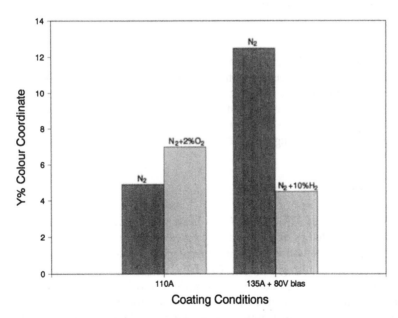

Figure 8.4 Variation of $y\%$ colour coordinate with impurity gases. Oxygen is more reactive than nitrogen increasing the colour coordinate whereas hydrogen depletes nitrogen in the film and reduces the colour coordinate.

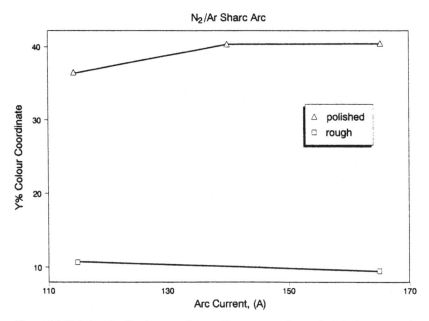

Figure 8.5 Variation of $y\%$ colour coordinate with arc current for rough (rolled) and smooth (polished) surfaces. Note that surface roughness plays a more significant role than deposition parameters in determining the surface colour.

wavelength of visible light increases scattering and absorption and, hence, leads to the darker coatings with high $y\%$ colour coordinates. Thus, the colour coordinates have a dual role. They may be used to monitor production runs, detecting contaminants, poor surface finish, or unsatisfactory cleaning of the surfaces. They may also be used to study microstructural changes and even changes in the electronic bands.

8.6 CONCLUSION

A full range of TiN coatings has been made using a titanium Sharc source. XRD studies of the films have revealed only the fcc δ-phase structure. A plot of the lattice constant, normal to the surface, shows that as the concentration of nitrogen in fcc TiN falls, the lattice constant increases. This is contrary to the equilibrium behaviour in α-Ti, ϵ-Ti$_2$N, or δ-TiN. In part, this may be attributed to stresses for nonstoichiometric films. As a metal, Ti atoms will have unit sticking coefficient independent of incident lattice site. Thus, in the presence of excess Ti, the structure would be expected to grow in a more disordered and, hence, more stressed form.

Nitrogen deficient films should also be more reactive to environmental oxygen. Since TiO has a smaller lattice constant, the presence of oxygen should be expected to reduce the lattice constant. However, the opposite behaviour was observed indicating increased stress levels associated with the different sized atoms.

8.7 REFERENCES

1 Jewsbury, P., Turley, N. and Ryan, N. "Status Report on Surface Engineering," TTCP report PTP-1/14/90, 1990.

2 Jewsbury, P. and Ramalingam, S. "Thin Film Deposition of TiN by Vacuum Arc Technology," in preparation.

3 Jewsbury, P., Ramalingam, S. and Chang, R. F. "True Surface Engineering: Composition Control in Deposits using Arc Technology" in *Engineered Materials for Advanced Friction and Wear Applications* F. Smidt, ed., ASM, Gaithersburg, 1988.

4 Jewsbury, P., Doyle, E. D., Lambrines, P., Downes, I. W. and Healy, G. "Design and Study of a Novel PVD Arc Source," in preparation.

5 Moore, P. G. "Arc Deposition of TiN," MRL technical report, in preparation.

6 Etchessahar, E., Bars, J. P. and Debuigne, J. "The TiN System: Equilibrium between the δ, ϵ and α Phases and the Conditions of Formation of the Lobier and Marcon Metastable Phase," *J. Less Common Metals,* 134:123–139, 1987.

7 Wyszecki, G. and Stiles, W. S. *Color Science: Concepts & Methods, Quantitative Data and Formulae, 2nd Edition,* J. Wiley, New York, 1982.

CHAPTER 9

Developments in Plasma-Assisted Physical Vapour Deposition

A. MATTHEWS[1]
K. S. FANCEY[1]
A. S. JAMES[1]
A. LEYLAND[1]

9.1 INTRODUCTION

THE benefits achievable by depositing PVD coatings in a plasma were first brought to the world's attention by Berghaus [1], but it was Mattox [2] who coined the term *ion plating* and initiated an upsurge in this technology, which has accelerated over the past decade. The expansion has been largely fueled by process developments that have enhanced the degree of ionisation achievable [3]. These developments, whilst considerably improving the quality and range of coatings that can be produced, have also increased the complexity of the processes. Research at Hull University has been geared towards improving understanding of the mechanisms occurring and exploiting the potential of the processes to produce new coatings. This chapter traces some of these studies under four headings:

(1) Fundamental Process Studies
(2) Plasma Processing
(3) Thermal Barrier Coatings
(4) Hard Coating Developments

9.2 FUNDAMENTAL PROCESS STUDIES

It has been claimed that, in some cases, commercial exploitation of PAPVD methods preceded a full scientific understanding of the processes.

[1]Research Centre in Surface Engineering, Department of Engineering Design and Manufacture, University of Hull, Hull HU6 7RX, UK.

It is true that it is only in recent years that it has been possible to explain fully why many of these processes are as effective as they are. This has come about through experimental plasma studies and theoretical analyses of deposition systems.

The earliest PAPVD ion plating processes utilized a simple Dc diode configuration with the surface to be coated as the cathode. Under the pressure range used, the chamber is filled by a "negative glow" (see Figure 9.1) [4]. Most of the voltage is dropped across a narrow region near to the sample, known as the cathode fall distance or sheath thickness L. Ions arriving at the cathode are neutralized close to the surface by a potential emission mechanism, which leads to the ejection of "secondary" electrons, typically with energies of 1–4 eV. These secondary electrons are accelerated across the cathode fall, increasing their energy. They are then known as primary electrons and are responsible for initiating further ionisation. The glow discharge is thus self-sustaining.

The cathode fall distance is extremely important, since it influences the energy distribution of ions arriving at the cathode and the uniformity of bombardment. Depending on the mean free path for charge exchange (λ), ions passing across the cathode fall may undergo collisions, transferring their charge to neutrals and leading to a combination of both ions and neu-

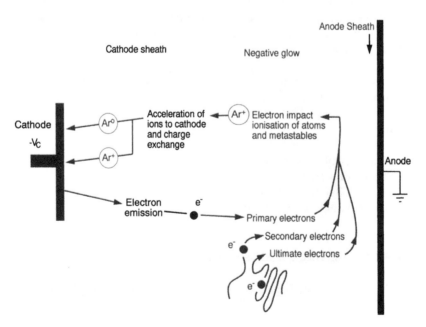

Figure 9.1 A low pressure DC diode argon glow discharge.

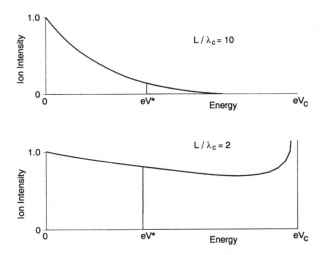

Figure 9.2 The influence of the L/λ value on the ion energy distribution.

trals arriving at the surface with a distribution of energies. An expression to allow the prediction of the ion energy spectrum has been presented by Rickards [5], based on work by Davis and Vanderslice [6]:

$$\frac{dN}{dE} = \frac{N}{m'} \frac{L}{\lambda} (1 - E)^{(1/m'-1)} \exp[-L/\lambda + L/\lambda(1 - E)^{1/m'}]$$

where E defines the ion energy relative to the maximum and N is the number of ions entering the cathode fall from the negative glow, and m' defines the field. In this equation there is a need to know L and λ to predict the ion energy distribution. The former can be found using the free-fall Child-Langmuir equation:

$$L = (4/9 \; \epsilon_0/J)^{1/2}(2q/m)^{1/4}V^{3/4}$$

where ϵ_0 = permittivity of free space, q = ionic charge, m = ion mass, and V = potential drip across L. Fancey and Matthews [7] showed that this allows L to be predicted for a wide range of process layouts. λ can be estimated from $\lambda = 1/n\sigma$ where σ is the charge exchange cross section and n the gas number density. It has been shown [8] that σ can be estimated for argon from available data and n can be evaluated from the discharge pressure. It is thus possible to calculate the cathode fall distance and the ion and neutral energy distribution. Figure 9.2 shows how the ion energy distribution is influenced by the L/λ value [9].

Clearly, if a minimum ion energy eV* is required in order to achieve some desired effect at the sample surface, then systems with the second distribution will be more effective. This corresponds to a low L/λ value, which in turn signifies the need for intensification of ionisation, in order to increase the current density to the sample. This has been interpreted as a need to increase the ionisation efficiency at the sample surface I_{ef}, where

$$I_{ef} = \frac{Ni \times 100\%}{N_{np}}$$

Ni is the number of arriving ions per cm² per sec, and N_{np} is the number of atomic bombardments per cm² per sec. Figure 9.3 shows how I_{ef} varies with pressure, at a given current density [11].

By consideration of existing commercial systems, it has been demonstrated that optimization of ion plating processes is achieved by enhancing the isolation; i.e., increasing I_{ef}, decreasing L, improving the ion energy spectrum, and increasing the ion current carried by metal species. All successful ion plating systems incorporate some form of ionisation enhancement, sometimes generated by the vapour source, but usually involving the

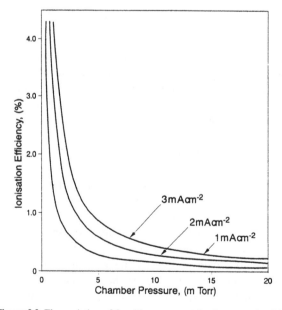

Figure 9.3 The variation of I_{ef} with pressure at fixed current densities.

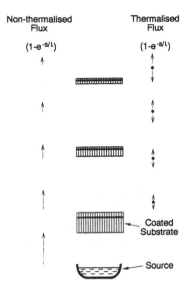

Figure 9.4 Basic principles of the thickness uniformity model.

increased availability of ionising electrons in the plasma [10,11]. This observation was originally made for evaporative systems but has now been extended to magnetron sputtering, where it can be achieved by "unbalancing" the magnetron [12].

One of the most important variables, regardless of the process used, is the substrate temperature. It has been shown that this can be predicted using models devised by Matthews [13,14]. It is important to note that, in enhanced discharge systems, the dominant heat source is the ion and energetic neutral bombardment. Thus, when nonuniform ion fluxes are present, the heating effect can vary throughout the chamber. This can occur on samples having different radii. Also, recent work [15–17] has shown that, in enhanced plasmas, there is a directionality to the bombardment, which also influences heating and sputter rates.

Another process variable is the thickness distribution. This can be influenced by bombardment and also by the distribution of the depositing material and samples throughout the chamber. It has been shown [18] that this can be predicted in terms of a front to back thickness ratio, R, by considering the "thermalized" and "nonthermalized" vapour fluxes within the chamber. Figures 9.4 and 9.5 illustrate this approach, which shows that

$$R = \frac{1 + \exp(-s/l)}{1 - \exp(-s/l)} = \coth\,(s/2l)$$

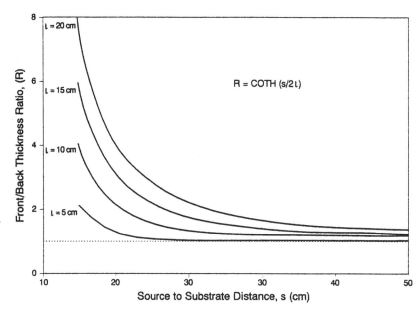

Figure 9.5 Variation of thickness uniformity with source to substrate distance (for different *l* values), as predicted by the model.

where *s* is the source to substrate distance and *l* is an associated mean free path. Further details of the model are given in References [19] and [20]. An interesting observation that can be made is the experimental value for *l* exceeds that which would be expected. There could be many reasons for this, but one is that the vapour moves as clustered atoms rather than single atoms. Furthermore, experimental validation of the model suggests that the vapour emanates from a "virtual" source at some distance above the actual source. This can be considered as a viscous cloud, which of course could be a region where clustering could occur. This cluster theory was given increased credibility when studies were made of the cathode sheath thickness *L* under evaporation conditions. Thus experiments have shown [21,22] that *L* is much less than that predicted by the Child-Langmuir equation if singly charged atoms are assumed to be the ionised species. Indeed, the model is found to apply only if very small charge to mass ratios are assumed – again, indicating the presence of singly charged clusters of metal atoms. This observation is potentially of major significance in the study of nucleation and growth mechanisms for films, the role of ions and neutrals, and a number of other process parameters and is now actively being researched further.

9.3 PLASMA PROCESSING

In view of our this laboratory's interest in plasma-assisted PVD technology, it is perhaps not surprising that it was chosen to investigate the role of "enhanced" plasma systems in surface modification. Over the past two decades, plasma (or ion) nitriding (a process first conceived during the early 1930s [23]) has become an accepted alternative to gas nitriding. However, it was felt that improvements to the process could be achieved by utilising the knowledge gained about plasma fundamentals and especially by operating the system at lower pressure. Comparative studies on various low pressure dc triode, RF (radio frequency) pulsed plasma systems were thus performed [24,25].

Using optical emission spectroscopy, the spatial distribution of species perpendicular to the cathode has been studied and the dominant ionic species (N_2^+ or N^+) for various systems determined [22,25]. Thus it has been possible to correlate this with nitriding performance and thereby optimise a plasma nitriding system that utilises a radio frequency plasma. The explanation for the improved hardness profiles achieved by this method appears to lie with the high ion energies produced at low pressure by the low frequency RF technique employed. Furthermore, in line with the work reported in Section 9.2, sheath thicknesses are small, with corresponding improvements in coverage uniformity and re-entrant penetration. Figure 9.6 shows the improved hardness profiles with depth that can be achieved with RF nitriding compared to low pressure Dc triode techniques. A major benefit of the low-pressure nitriding methods so-developed is their ability to control white layer growth on ferrous materials. This is particularly important when post-coating of components is required. The samples do not need to be removed from the chamber and prefinished prior to coating for acceptable levels of adhesion to be achieved.

9.4 THERMAL BARRIER COATINGS

There are many heat engine developments that are surface-constrained; i.e., existing materials cannot function satisfactorily at the desired surface operating temperatures. If these could be achieved, there would be massive savings in fuel economy and in increased power output.

In order to be effective, a thermal barrier coating (TBC) system needs to provide good thermal insulation, resistance to thermal shock, and hot corrosion/oxidation and particulate erosion resistance, in addition to having an aerodynamically efficient surface. These demands are clearly very

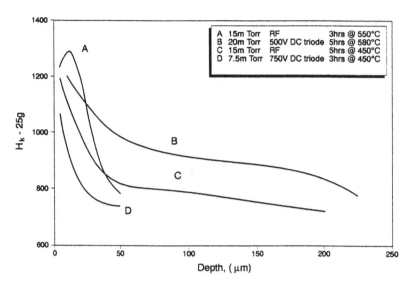

Figure 9.6 Hardness depth profiles on H13 steel produced by different nitrogen discharge conditions.

challenging, but plasma-assisted PVD coatings developed at Hull are proving effective in meeting them.

Zirconia-based TBCs have been used in the combustor and after-burner regions of many aero engines since the mid-1970s, but in today's commercial market, it is almost exclusively the static components that are coated, due largely to the more severe stresses that the coatings must endure on rotating components such as the turbine blades. Essentially, the problem becomes one of adhesion of the TBC to the substrate under the relevant conditions.

Commercial TBCs have been produced by the plasma-spraying technique; however, such coatings require a rough substrate to maximize adhesion, and the eventual coating also has a rough surface. These rough surfaces are detrimental to the interfacial strength, oxidation resistance, and aerodynamic efficiency. In addition, accurate thickness control is difficult to achieve on complex components.

More recently, electron beam physical vapour deposition (EBPVD) has emerged with some success as an alternative deposition process [26,27]. This involves the evaporation of a source material with a high power electron beam gun and is essentially a line-of-sight process. A large increase in thermal cycle durability has been achieved by use of EBPVD rather than plasma spraying in burner rig tests. The reason for the improved durability is attributed to the segmented microstructure formed in the EBPVD process, which improves the strain tolerance and the thermal shock resistance. In addition, the smoother surfaces achieved in EBPVD have been shown to be more oxidation and erosion resistant. A common composition of the zirconia coatings for thermal durability is ZrO_2–6 to 8 wt % Y_2O_3. This material is known as a partially yttria stabilized zirconia (PYSZ).

Since 1982, the practicability of depositing TBCs such as PYSZ under plasma-assisted PVD conditions has been investigated (and this work has been highly successful) confirming the view that the improved adhesion and structure control achievable by this method can enhance coating performance considerably [28–32,34]. Indeed, the optimum structure for TBCs in thermal cycling trials was found to be a mixed structure consisting of a thin dense plasma-assisted deposit near to the substrate, followed by a thicker porous gas-evaporated layer. Such a controlled change in the structure would not be easy to produce by any method other than plasma-assisted PVD (Figure 9.7).

9.5 HARD COATING DEVELOPMENTS

In addition a major effort has been devoted to the development of new hard coatings, usually for tribological applications. This work has included, e.g., the further development of TiN coatings [35–40], TiC [41], TiO_2 [42], Al_2O_3 [41], TiBN and TiAlN [43–47], and diamond-like coatings [48–54]. Many materials pairings have been evaluated on a pin on disc machine, which conforms to the VAMAS European Standard and includes humidity and temperature control. Tests are typically carried out unlubricated against a coated or uncoated hardened steel spherical pin—the disc usually being coated hardened steel. In Reference [42], it was shown that several modes of friction and wear performance can be achieved by using different coatings. In that work, three main types were reported: (1) low friction and low pin and disc wear (TiN or TiC coated pin against a TiN coated disc); (2) intermediate friction and low pin and disc wear (TiC or TiO coated pin against an uncoated steel disc); and (3) high friction and high disc wear TiN coated or uncoated pin rubbing against an uncoated steel disc).

More recent work [43,44] has examined the pin on disc test in greater detail and, in particular, the variability in test performance. The contact tribochemistry was found to be a critical factor in determining the friction and wear behaviour. Thus, advanced multi-phase ceramics have great potential in controlling friction and wear (e.g., TiBN films). Similarly doped or mixed-phase carbon coatings containing boron and nitrogen provide benefits over carbon alone [50]. This is probably in part due to chemical effects at the sliding interface and also due to toughening by the second or third phase additions. The coatings mentioned so far in this section

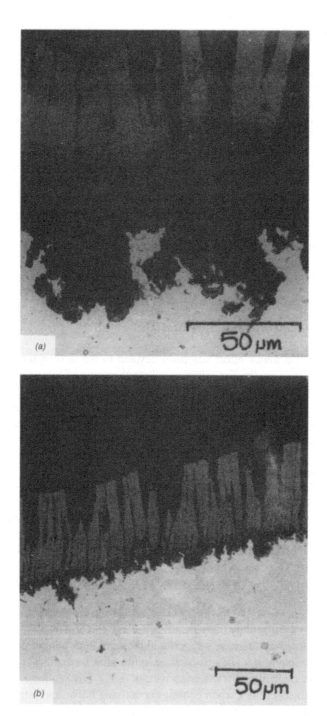

Figure 9.7 Corrosion of nimonic material under the TBC, after burner rig trials: (a) single layer TBC; (b) double layer TBC.

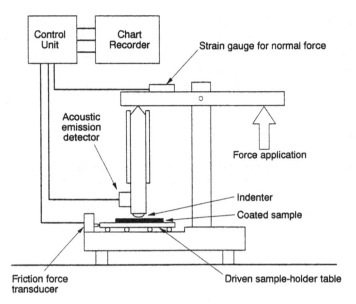

Figure 9.8 The scratch test.

were deposited by a thermionically assisted triode ion plating method, in which the coating material was evaporated using an electron beam gun. CVD and fast atom beam methods have also been utilised for the deposition of hard carbon films. These studies have revealed that the impact angle of the beam on the substrate critically influences the nature of the film, including the optical transparency and hardness [52].

The laboratory has also been particularly involved in the development and assessment of standardised techniques to evaluate coatings [55–58] and in the pursuit of applications [59–61]. Of primary importance is the need to be able to evaluate adhesion. Whilst no fully quantitative and nondestructive method is presently available, it has been shown that the scratch test can provide a useful insight into modes of coating failure (Figure 9.8).

9.6 CONCLUSIONS

The developments reported here are aimed towards process developments that will ultimately lead to the improvement of advanced coatings and their wider adaption in industry. With the latter point in mind, computer expert systems are also being developed that should aid the design engineer in selecting coatings for particular applications [62–64], in conjunction with major European companies as part of an EC funded project. Also, the further development of the research into plasma-assisted PVD is being carried out in conjunction with the Polytechnics of Newcastle and Sheffield City. The latter partner institutions are participating with Hull University in a major surface engineering initiative in the UK.

9.7 ACKNOWLEDGEMENTS

The work reported here has been financed by industrial supporters, the SERC, and the EC BRITE/EURAM scheme. We gratefully acknowledge that support and the help of our coworkers in the Research Centre in Surface Engineering, and at Sheffield City and Newcastle Polytechnics.

9.8 REFERENCES

1 Berghaus, B. UK Patent 510.933, 1939.

2 Mattox, D. M. *Electrochem. Technol.*, 2:295, 1964.

3 Matthews, A. *J. Vac. Sci.-Technol.*, A3: 2345, 1985.

4 Fancey, K. S. and Matthews, A. In *Advanced Surface Coatings*, D. R. Rickerby and A. Matthews, eds., Blackie, 1991.

5 Rickards, J. *Vacuum*, 34:559, 1984.

6 Davies, W. D. and Vanderslice, T. A. *Phys. Rev.*, 131:219, 1963.

7 Fancey, K. S. and Matthews, A. *Proc. 6th Int. Conf. on Ion and Plasma Assisted Techniques*, PAT 87, Brighton, CEP Edinburgh, 1987.

8 Fancey, K. S. and Matthews, A. *Surf. Coat. Technol.*, 33:17, 1987.

9 Chapman, B. N. *Glow Discharge Processes*,. Wiley, New York, 1980.

10 Matthews, A. In *Physics and Chemistry of Protective Coatings*, W. D. Sproul, J. E. Greene and J. A. Thornton, eds., American Institute of Physics, Conference Proceedings No. 149, AIP, New York, 1986.

11 Matthews, A. *Surface Engineering*, 1:93, 1985.

12 Robinson, P. A. and Matthews, A. *Proc. Int. Conf. on Metallurgical Coatings*, ICMC 90, San Diego, USA, 1990. *Surf. Coat. Technol.*, 43/44:288-298, 1990.

13 Matthews, A. *Vacuum TAIP*, 32:311, 1982.

14 Matthews, A. and Gethin, D. T. *Thin Solid Films*, 117:261, 1984.

15 Fancey, K. S. and Matthews, A. *IEEE Transactions on Plasma Science*, 18:6, 1990.

16 Fancey, K. S. and Matthews, A. *Proc. Int. Conf. on Metallurgical Coatings*, ICMC 90, San Diego, USA, 1990.

17 James, A. S., Fancey, K. S. and Matthews, A. *Proc. 2nd Int. Conf. on Plasma Surface Engineering*, Garmisch Partenkirche, FRG, DGM Oberursel, 1990.

18 Fancey, K. S. and Matthews, A. *Surf. Coat. Technol.*, 36:233, 1988.

19 Fancey, K. S. and Matthews, A. *Proc. 1st Int. Conf. on Plasma Surface Engineering*, Garmisch Partenkirchen, FRG, DGM, Oberursel, 1989.

20 Fancey, K. S., Robinson, P. A., Leyland, A., James, A. S. and Matthews, A. *Proc. 2nd Int. Conf. on Plasma Surface Engineering,* Garmisch Partenkirchen, FRG, DGM, Oberursel, 1990. *Materials Science and Technology,* A140:576, 1991.

21 Fancey, K. S. and Matthews, A. *Appl. Phys. Lett.,* 55:834, 1989.

22 Fancey, K. S. and Matthews, A. 1990. *Proc. Int. Vacuum Congress,* Cologne, Sept. 1989. *Vacuum,* 41:2196, 1990.

23 Berghaus, B. German Patent DRP 668639, 1932.

24 Leyland, A., Fancey, K. S., Bell, T., Kwon, S. C., Park, M.-J. and Matthews, A. *Proc. Plasma Surface Engineering Conf. PSE 88,* Garmisch Partenkirchen, Germany.

25 Leyland, A., Fancey, K. S., James, A. S. and Matthews, A. *Surface and Coatings Technology,* 41:295, 1990.

26 Demaray, R. F., Fairbanks, R. W. and Boone, D. H. ASME Rep 82-GT-264, 1982.

27 Sumner, I. E. and Ruckle, D. L. AIAA Paper 80-1193, 1980.

28 Fancey, K. S. and Matthews, A. *Proceedings of the International Conference on Metallurgical Coatings, ICMC 86,* San Diego, USA, 1986. *I. Vac. Sci. Technol.,* A4(6):2656, 1986.

29 Fancey, K. S., James, A. S. and Matthews, A. *Proceedings of the International Conference: Engineering and Surface,* Institute of Metals, London, 1986.

30 James, A. S., Fancey, K. S. and Matthews, A. *Proceedings of the International Symposium on Trends and New Applications in Thin Films,* Strasbourg, Societe Francaise du Vide, Paris, 1987.

31 James, A. S., Fancey, K. S. and Matthews, A. *Proceedings of the International Conference on Metallurgical Coatings, ICMC 87,* San Diego, USA, 1987. *Surface and Coatings Technology,* 32:377–387, 1987.

32 James, A. S. and Matthews, A. *Proceedings of the Plasma Surface Engineering Conference, PSE 88,* Garmisch Partenkirchen, West Germany, 1988.

33 James, A. S. and Matthews, A. *Proceedings of the International Conference on Metallurgical Coatings, ICMC 89,* San Diego, USA, 1989. *Surface and Coatings Technology,* 41:305, 1990.

34 James, A. S. and Matthews, A. *Proc. Int. Conf. on Metallurgical Coatings ICMC 90,* San Diego, USA, 1990. *Surface and Coatings Technology,* 43/44:445, 1990.

35 Matthews, A. and Sundquist, H. A. *Proceedings of the International Ion Engineering Congress, ISIAT 83,* Kyoto, Japan, IEE, 1983.

36 Matthews, A., Valli, J., Makela, U. and Murawa, V. *Proceedings of the International Conference on Metallurgical Coatings,* Los Angeles, USA, 1985. *Journal of Vacuum Science and Technology,* A3:2411, 1985.

37 Matthews, A. and Lefkow, A. R. *Proceedings of the 6th International Conference on Thin Films, ICTF6,* Stockholm Sweden, 1984. *Thin Solid Films,* 126:283, 1985.

38 Sundquist, H. A., Matthews, A. and Valli, J. *Proceedings of the 4th International Tribology Symposium — Eurotrib 85,* Lyon, France, published by the International Tribology Council, 1985.

39 Matthews, A. and Murawa, V. *The Chartered Mechanical Engineer,* 32:31, 1985.

40 Matthews, A. *Engineering,* Technical File, December 1987.

41 Matthews, A. and Valli, J. *Proceedings of the Surtec 85 Conference — Interfinish,* Berlin, West Germany, published by Verein Deutscher Ingeniere (VDI), Dusseldorf, 1985.

42 Matthews, A., James, A. S., Leyland, A., Fancey, K. S., Valli, J. and Stainsby, J. A. *Proceedings of Tribology 87*, London, Institution of Mechanical Engineers, London, 1987.

43 Ronkainen, H., Varjus, S., Holmberg, K., Fancey, K. S., Pace, A. R., Matthews, A., Matthes, B. and Broszeit, E. *Proceedings of the 16th Leeds-Lyon Tribology Symposium*, Lyon, France, Elsevier Sequoia, 1989.

44 Ronkainen, H., Homberg, K., Fancey, K. S., Matthews, A., Matthes, B. and Broszeit, E. *Proc. Int. Conf. on Metallurgical Coatings, ICMC 90*, San Diego, USA, 1990. *Surface and Coatings Technology*, 43/44:888–897, 1990.

45 Park, M. J., Leyland, A. and Matthews, A. *Proc. Int. Conf. on Metallurgical Coatings, ICMC 90*, San Diego, USA, 1990. *Surface and Coatings Technology*, 43/44:481, 1990.

46 Ronkainen, H., Leyland, A. and Matthews, A. *Proc. 2nd Int. Conf. on Plasma Surface Engineering, PSE 90*, Garmisch Partenkirchen, Germany, 1990. *Materials Science and Engineering*, A140:722, 1991.

47 Ronkainen, H., Nieminen, I., Homberg, K., Leyland, A., Fancey, K. S., Matthews, A., Matthes, B. and Broszeit, E. *Proc. 2nd Int. Conf. on Plasma Surface Engineering, PSE 90*, Garmisch Partenkirchen, Germany, 1990. *Materials Science and Engineering*, A140:602, 1992.

48 Fitzgerald, A. G., Simpson, M., Dederski, G. A., Moir, P. A., Tither, D. and Matthews, A. *Carbon*, 26:229–234, 1988.

49 Tither, D., Matthews, A., Fitzgerald, A. G., Storey, R. E., Henderson, A. E., Moire, P. A., Dynes, T. J., Bower, D. I., Lewis, E. L. V., Doughty, G. and Foster, W. *Carbon*, 27(4):655, 1989.

50 Matthews, A., Tither, D. and Holiday, P. *Proceedings of the 1989 European Tribology Conference, EUROTRIB 89*, Helsinki, Finland, International Tribology Council, London, 1989.

51 Dehbi-Alaoui, A., James, A. S. and Matthews, A. *Proc. Int. Conf. on Metallurgical Coatings, ICMC 90*, San Deigo, USA, 1990. *Surface and Coatings Technology*, 43/44:88, 1990.

52 Dehbi-Alaoui, A., Matthews, A. and Franks, J. *Proc. Diamond Films 90*, Crans Montana, Switzerland, 1990. *Surface and Coatings Technology*, 47:722, 1991.

53 Dehbi-Alaoui, A., Holiday, P. and Matthews, A. *Proc. Diamond Films 90*, Crans Montanta, Switzerland, 1990. *Surface and Coatings Technology*, 47:327, 1991.

54 Holiday, P., Dehbi-Alaoui, A. and Matthews, A. *Proc. Diamond Films 90*, Crans Montana, Switzerland, 1990. *Surface and Coatings Technology*, 47:315, 1991.

55 Valli, J., Makela, U. and Matthews, A. *Surface Engineering*, 2:49, 1986.

56 Matthews, A. and Valli, J. *Proceedings of the IFHT International Seminar on Plasma Heat Treatment*, Senlis, France, International Federation for the Heat Treatment of Materials, Paris 1987.

57 Matthews, A. *Proceedings of the Surfaces, Materiaux, Technologie Conference*, Lyon, 1988. Le Vide, les Couches Minces, No Special, SFV, Paris 1988.

58 Bull, S. J., Rickerby, D. S., Matthews, A., Leyland, A., Pace, A. R. and Valli, J. *Proceedings of the International Conference on Metallurgical Coatings, ICMC 88*, San Diego, USA 1988. *Surface and Coatings Technology*, 36:503, 1988.

59 Matthews, A. *Proceedings of the 2nd International Conference on Physical Vapour Deposition, PVD 86*, Darmstadt, West Germany, 1986.

60 Matthews, A. *Advanced Materials and Manufacturing Processes*, 3:91–105, 1988.

61 Matthews, A. *Proceedings of the 1st International Conference on Surface Modification Technologies,* Pheonix, Arizona, The Metallurgical Society/AIME, Warrandale, 1988.

62 Matthews, A. *Proceedings of the Institute of Metals Autumn Meeting—A Cutting Edge for the 1990s,* Sheffield, Institute of Metals, London, 1989.

63 Syan, C. S., Matthews, A. and Swift, K. G. *Surface Engineering,* 2:249, 1986.

64 Syan, C. S., Matthews, A. and Swift, K. G. *Proceedings of the International Conference on Metallurgical Coatings, ICMC 87,* San Diego, USA, 1987. *Surface and Coatings Technology,* 33:105–115, 1987.

65 Matthews, A. *Proceedings of the International Conference on Materials and Engineering Design,* London, The Institute of Metals, London, 1988.

CHAPTER 10

Plasma Nitriding

N. L. LOH[1]

10.1 INTRODUCTION

NITRIDING is a ferritic thermochemical treatment, which introduces atomic nitrogen into the ferrite phase of ferrous materials. It is a case-hardening technique operating in the temperature range of 500°C–590°C, 590°C being the eutectoid temperature in the binary iron-nitrogen system. The process was first used in the 1920s, and since then, its industrial application has remained largely unchanged. Basically, there are three conventional methods of nitriding — gas, salt-bath, and powder nitriding. The case depth and hardnesses obtained are primarily dependent on the treatment time and temperature, the nitrogen activity of the nitriding medium and the alloying content of the ferrous materials.

Industrially, the gas-nitriding technique is most commonly practiced to case-harden ferrous materials. This is basically a thermally activated process. Anhydrous ammonia dissociates to produce atomic nitrogen upon contact with workpieces heated to 500°C–580°C. The atomic nitrogen reacts with alloying elements such as Al and Cr to form the desirable nitrides. Some convert to molecular nitrogen, which is inert and useless for nitriding purposes. Effective nitriding, therefore, depends on producing atomic nitrogen at the surface of the workpiece for immediate chemical reaction.

The rather limited temperature range of 500°C–580°C, however, is a disadvantage. Futhermore, gas nitriding tends to produce brittle compound layers consisting of both the cph-ϵ and fcc-γ phases. By virtue of

[1]Nanyang Technological Institute, School of Mechanical and Production Engineering, Nanyang Avenue, Singapore 2263.

their structural differences and misfit, high inherent stresses can be induced to initiate microcracks. For this reason, they are often removed before the nitrided workpiece is put to service. This is particularly important where the compound layer measures more than 25 microns. Mechanical lapping is often employed to remove the compound layer; and it has been established that the cost of lapping per unit can be as high as the nitriding treatment itself.

10.2 PLASMA NITRIDING

Plasma nitriding operates under the glow discharge conditions. Accordingly, it is also known as glow discharge nitriding or simply ionitriding. The process was first patented in the early 1930s by a Swiss engineer, Bernard Berghaus. It utilizes the highly active and reactive plasma state to introduce ionic nitrogen species into ferrous materials. Theoretically, the process can impregnate any interstitial elements such as N, C, and B into any electrically conductive material if the elements can be processed in a gaseous state.

10.2.1 GLOW DISCHARGE FUNDAMENTALS

Figure 10.1 shows the voltage-current characteristics of the various types of gas discharges. Plasma nitriding operates in the region of the abnormal glow discharge, where the voltage-current relationship bears a single-valued function. Care must be observed to ensure that the power used does not exceed point G, or damages will be incurred in the arc discharge region. Indeed, it was technical difficulties associated with the maintenance of the unstable, abnormal glow discharge that had deterred its initial industrial application. Today, however, these problems are solved largely by the employment of power electronics and microprocessor control.

A gas discharge is generated by establishing a voltage across two electrodes in soft vacuum. Figure 10.2 shows schematically the electrostatic field strength E and the voltage V, across the electrodes, together with the density distribution of the ions n_i and electrons n_e. In a glow discharge, the plasma state is in non-LTE (local thermodynamic equilibrium) with a typical density of 10^{16}–10^{20} particles per m^3 [1].

Under the glow discharge condition, electrons leave the cathodic surface by secondary electron emission and speed through the positive column towards the anode. In the course of travel, they collide with the gaseous atoms and molecules, converting them to ions. By virtue of their positive charge, the ions accelerate across the electrostatic field towards the cathode, thereby promoting further secondary electron emission. For a

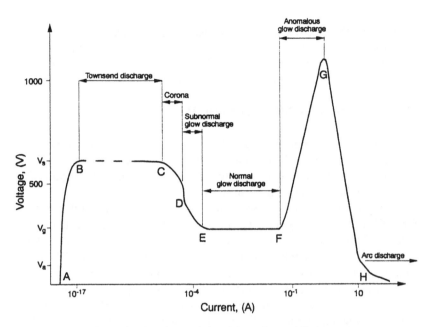

Figure 10.1 V-I characteristics of the various gas discharges.

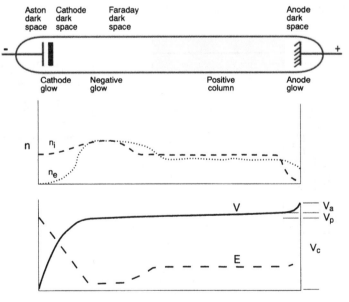

Figure 10.2 Schematic diagram of the glow discharge features.

stable glow, the electrostatic field intensity and gas density must be such as to bring about a self-sustaining state. This is achieved when the cathodic secondary electron emission and the electron-gas collision processes provide sufficient electrons to maintain the discharge.

10.2.2 PLASMA-NITRIDING MECHANISM

Figure 10.3 is a schematic diagram of the plasma-nitriding system. It is comprised essentially of four units: the furnace proper, the power supply and control unit, the pump, and the gas distribution unit. Figure 10.4 shows an industrial installation in the UK.

With pressure down to less than 1 mbar, a potential difference of 600–800 volts can be established across the cathodically connected workpiece and the grounded furnace wall. This creates an electrostatic field within the furnace chamber and ionizes the furnace atmosphere. The electrostatic field covers the workpiece surfaces uniformly. This ensures uniform treatment of all exposed surfaces. Nitrogenous gas is then introduced as the nitriding medium. This may be a N_2-H_2 or NH_3-H_2 gas mixture.

Under glow discharge conditions, the nitrogenous gas is ionized. The nitrogen ions accelerate across the electrostatic field and bombard the cathodically connected workpiece. The bombarding process of the ions sputters and cleanses the surface effectively, removing any diffusion barriers on the workpiece surfaces, such as contaminants and oxides. As the

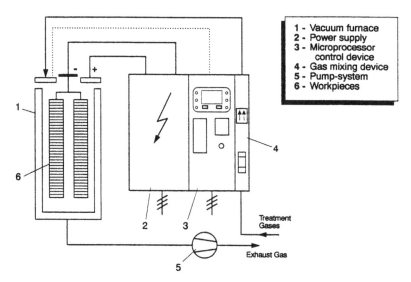

1 - Vacuum furnace
2 - Power supply
3 - Microprocessor control device
4 - Gas mixing device
5 - Pump-system
6 - Workpieces

Figure 10.3 Schematic diagram of the plasma-nitriding system.

Figure 10.4 150-W industrial installation in the UK.

ions come to rest, the liberation of their kinetic energies heats up the work-piece. Nitrogen occludes onto the cathodic surface and reacts to form nitrides.

Figure 10.5 shows a model of the plasma-nitriding mechanism. Dressler [2] and Edenhofer [3] postulated that the bombarding nitrogen reacts with iron to form FeN on the surface of the workpiece. FeN is thermo-dynamically unstable under the nitriding condition. Hence, this is reduced progressively to ϵ-$Fe_{2-3}N$ and γ-Fe_4N to form the white compound layer. Nitrogen also diffuses into the matrix as interstitial nitrogen in solid solution and alloy nitride precipitates.

10.2.3 MICROSTRUCTURES

By varying the appropriate processing parameters such as temperature, treatment time, gas composition, and pressure, plasma nitriding can pro-duce any one of the following three microstructures:

(1) A diffusion layer with a poly-phased compound layer of γ-Fe_4N and ϵ-$Fe_{2-3}N$

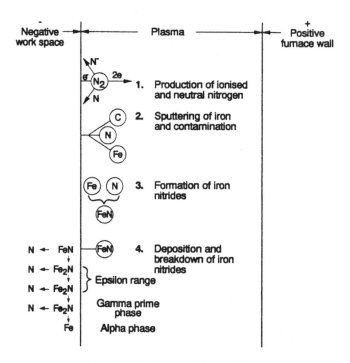

Figure 10.5 Mechanism of plasma nitriding.

(2) A diffusion layer with a monophased compound layer of γ-Fe$_4$N or ϵ-Fe$_{2-3}$N

(3) A diffusion layer without any compound layer

Figure 10.6 shows a typical microstructure of a plasma-nitrided steel. The compound layer develops when nitrogen reaches the workpiece surface at a rate faster than its diffusion into the matrix [4]. The nitrogen content at the surface of the workpiece is primarily a function of the gas composition and pressure. Its desirability depends principally on two factors: (1) the homogeneity of the nitride structure and (2) the thickness of the layer.

Generally, the mechanical properties are more superior if the compound layer is of monophased nitride. With a poly-phased compound layer of γ-Fe$_4$N and ϵ-Fe$_{2-3}$N, high inherent stresses can result at transition regions between the two different lattice structures. This can readily induce microcracks when the treated workpiece is subjected only to low nominal loads.

The monophased γ-Fe$_4$N is compact and wear-resistant. Hence, it is seldom ground off and the treated components are normally put to service without any further treatment. Owing to its narrow homogeneity range, the γ-Fe$_4$N tends to exhibit limited thickness.

The monophased ϵ-Fe$_{2-3}$N is known to improve the wear and corrosion resistances of cast iron [2]. Carbon stabilizes the ϵ-Fe$_{2-3}$N phase. For this reason, it is readily formed in alloy steels with >1% carbon.

It is generally accepted that the ductility of the layer will decrease with increasing thickness. Hence, the compound layer is normally kept to the minimum possible to meet service specification. With plasma nitriding, the thickness of the compound layer for a given material can be monitored by the rate of sputtering, the treatment time, temperature, and gas composition [5].

10.3 INDUSTRIAL APPLICATIONS

The industrial applications of plasma nitriding and plasma nitrocarburizing are fast gaining popularity. Often, they replace the conventional techniques of gas and salt-bath nitriding. In particular, salt-bath nitrocarburizing is becoming obsolete as regulations concerning environmental safety become more stringent. In Singapore, for example, the process of salt-bath nitrocarburizing is hardly ever used. Plasma nitriding and nitrocarburizing pose no environmental hazard. Along with their versatility and controllability, the processes are gaining inroads into the industrial scene.

Plasma heat treatment can handle the entire range of ferrous materi-

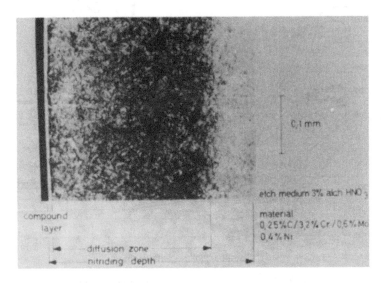

Figure 10.6 Typical microstructure of plasma-nitrided steel.

als—from cast irons and low carbon steels to alloy steels and stainless steels. Table 10.1 provides some examples of the industrial practices on the various grades of ferrous materials.

Perhaps it is fair to add that the superiority of plasma nitriding over those of the conventional methods lies in its capability to control accurately the growth and microstructure of the nitride layer. By controlling the temperature, treatment time, and gas composition, the thickness of the γ-Fe_4N compound layer can be closely monitored [5]. With 2.5% N_2–97.5% H_2 gas mixture, for example, plasma nitriding of 3% Cr-Mo steel at 480°C will not produce any compound layer at all, even with treatment time up to 50 hr. With alloy and stainless steels, >15% N_2 will produce either a thin γ layer or no compound layer at all.

Almost 80% of all application requirements, however, can be met with a microstructure consisting of a thin γ-Fe_4N compound layer with a diffusion depth of about 0.15 to 0.25 mm. The compound layer enhances wear and corrosion resistances (except for stainless steels) whilst the diffusion layer promotes fatigue resistance, particularly the bending and rotating bending fatigue [5]. Figure 10.7 shows some examples of microhardness traverse curves for the various grades of steels.

With cold-working tools, which are often subjected to high stresses and impact loads, a high core strength >62 HRC is often necessary to sup-

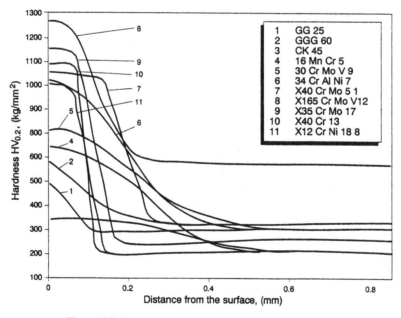

1	GG 25
2	GGG 60
3	CK 45
4	16 Mn Cr 5
5	30 Cr Mo V 9
6	34 Cr Al Ni 7
7	X40 Cr Mo 5 1
8	X165 Cr Mo V12
9	X35 Cr Mo 17
10	X40 Cr 13
11	X12 Cr Ni 18 8

Figure 10.7 Some typical microhardness curves of treated parts.

TABLE 10.1. Industrial Plasma Nitriding of the Various Grades of Ferrous Materials.

Type of Steels			Tensile Strength		Temperature Range		Surface Hardness	Nitriding Case Depth	
DIN	Material No.	AISI/SAE	(N/mm²)	KSI (1000 psi)	°C	°F	HVI	(mm)	(mil)
1) Structural Steels									
St 37-3	1.0116	~1020	370–450	55–65	550–580	1020–1080	200–350	0.3–0.8	12–32
St 52-3	1.0570	~1019	550–700	80–100	550–580	1020–1080	200–450	0.3–0.8	12–32
2) Carbon Steels									
C10	1.0301	1010	350–500	50–75	550–580	1020–1080	200–350	0.3–0.8	12–32
C45	1.0503	1045	670–820	90–120	550–580	1020–1080	300–500	0.3–0.8	12–32
3) Free Cutting Steels									
9S20K	1.0711	1212	350–550	50–80	550–580	1020–1080	200–400	0.3–0.8	12–32
9SMnPb28	1.0718	12L13	400–550	60–80	550–580	1020–1080	200–400	0.3–0.8	12–32
4) Cast Iron									
GG25	—	40B	≈250	≈35	530–570	980–1060	350–500	0.1	4
GGG60	—	80-55-06	≈600	≈85	530–570	980–1060	450–650	0.1–0.3	4–12
GTS55	—	—	≈550	≈80	520–560	970–1040	250–400	0.1	4
5) Powder Metal									
SINT	—	—	≥350	≥50	580	1080	260–350	0.1	4
6) Cementation Steels									
16MnCr5	1.7131	5115	550–700	80–100	520–550	970–1020	500–700	0.3–0.8	12–32
15CrNi6	1.5919	≈4320	550–700	80–100	520–550	970–1020	500–650	0.03–0.6	12–24

(continued)

165

TABLE 10.1. (continued).

Type of Steels			Tensile Strength		Temperature Range			Surface Hardness	Nitriding Case Depth	
DIN	Material No.	AISI/SAE	(N/mm²)	KSI (1000 psi)	°C	°F		HVI	(mm)	(mil)
7) Spring Steels										
67SiCr5	1.7103	—	1500–1600	218–233	≤420	≤790		700–800	≤0.1	≤4
8) Heat Treatable Steels										
42CrMo4	1.7225	4140	800–1100	120–160	500–550	930–1020		550–650	0.3–0.5	12–20
30CrNiMo8	1.6580	≈4340	800–1100	120–160	490–540	910–1000		600–700	0.3–0.5	12–20
30CrMoV9	1.7707	—	800–1100	120–160	490–540	910–1000		750–850	0.2–0.5	8–20
9) Nitriding Steels										
31CrMo2	1.8515	EN40B	900–1200	130–175	490–540	910–1000		750–900	0.2–0.5	8–20
34CrAlNi7	1.8550	A355	850–1100	125–160	520–550	970–1020		900–1000	0.2–0.5	8–20
10) Heat Resisting Steels										
14CrMoV69	1.7735	≈514	900–1100	125–160	490–540	910–1000		750–900	0.4–0.8	16–32
56NiCrMoV7	1.2714	L16	1100–1300	160–185	450–550	840–1020		550–700	0.2–0.5	8–20
11) Hot Working Steels										
X32CrMoV33	1.2365	1110	1400–1600	200–235	480–530	890–990		900–1100	0.1–0.3	4–12
40CrMoV51	1.2344	1113	1400–1600	200–235	480–530	890–990		900–1100	0.1–0.3	4–12
12) Cold Working Steels										
X100CrMoV51	1.2363	A2	1800–2200	260–320	480–510	890–950		800–1000	0.1–0.2	4–8
X155CrVMo122	1.2379	D2	≥2400	≥350	480–510	890–950		900–1300	≈0.1	≈4

166

TABLE 10.1. (continued).

Type of Steels			Tensile Strength		Temperature Range		Surface Hardness	Nitriding Case Depth	
DIN	Material No.	AISI/SAE	KSI (N/mm²)	(1000 psi)	°C	°F	HVI	(mm)	(mil)
13) High Speed Steels									
S6-5-2	1.3343	M2	≥2400	≥350	480–510	890–950	1000–1250	0.02–0.1	0.8–4
S18-0-1	1.3355	T1	≥2400	≥350	480–510	890–950	1000–1250	0.02–0.1	0.8–4
14) Managing Steels									
X2NiCoMo18 85	1.6359	—	≥2100	≥305	460	860	850–950	0.05–0.1	2–4
15) Stainless Steels									
X20Cr13	1.4021	420	650–950	95–140	540–570	1000–1060	850–1050	0.1–0.2	4–8
X35CrMol7	1.4122	≈434	800–950	120–140	550–580	1020–1080	950–1100	0.1–0.2	4–8
X5CrNiI89	1.4301	304	500–750	75–110	550–580	1020–1080	900–1200	0.05–0.1	2–4
16) Valve Steels									
X45CrSi93	1.4718	HNV3	900–1100	130–160	530–560	990–1040	600–1000	0.02–0.1	0.8–4
X55CrMnNiN208	1.4875	—	900–1150	130–165	550–580	1020–1080	700–1000	0.02–0.1	0.8–4

port the nitride layer. To retain a high core hardness, the nitriding temperature must be 30–50°C below the tempering temperature. Plasma nitriding can operate at temperatures as low as 350°C although the reaction kinetics are slow [6]. This allows a much lower tempering temperature, which produces a high core hardness. Industrial experience has shown that the maximum surface hardness for medium alloy steels is achieved in the temperature range of 450–480°C. With additional control of a slower cooling rate, low temperature plasma nitriding minimizes shape distortion and growth. Surface roughening is also observed to be minimal.

Over the past decade or so, plasma nitriding has gained much inroad into the industrial scene. It is used to case-harden components such as gears, crankshafts, extruder screws, hydraulic rams, valves, and even ball-point pen balls. An automotive company has begun to adopt the plasma-nitriding process into its production line [7]. Ford Motor Co has verified that plasma nitriding improves the wear lives of moulds by a factor of seven [8]. This led to further specifications of the ionitriding process for components such as injection-moulding screws, extrusion screws, valves, end plates, and the like.

In Singapore, the DOXON Engineering Pte Ltd was set up to plasma nitride components that are CNC-machined to high precision. To date, the commercial application of plasma nitriding is still very much in its introductory phase, and a lot more can be done to enhance its knowledge and application among the industries.

10.4 ADVANTAGES AND DISADVANTAGES OF PLASMA NITRIDING

The advantages of plasma nitriding over those of the conventional methods are well documented. These can be classified under four headings: metallurgical, technical, economical, and environmental.

By virtue of the electrostatic field, which envelops the workpiece surfaces uniformly, the nitride layer is more uniform and consistent. Composition and thickness of the compound layer can also be monitored. By and large, the microstructure is more ductile, and mechanical properties such as fatigue and wear are improved. Ionitrided parts are often reported to have longer service lives.

Technically, the plasma-nitriding process can operate over a much wider temperature range: from 400–580°C. Uniform and effective temperature control also ensures minimum shape distortion. In addition, there are no depassivation problems, and stainless steels and tool steels that are difficult to nitride by the conventional methods can be ionitrided.

As a result of ionic bombardment, the ionitriding effect begins very early even during the heating up process. Where small case depths are concerned, this would mean an effectively shortened treatment time. The process operates in soft vacuum. Gas consumption is therefore very low, and there is also a minimum loss of heat. In addition, no lapping or polishing is required in most cases. Rejection rates are extremely low. Thus, the process is economically very attractive.

Plasma nitriding poses no pollution problem, no explosion risk, nor any environmental hazards or effluent disposal problems. It is environmentally friendly.

Perhaps its major disadvantage lies in the initial high capital cost as compared to conventional nitriding furnaces. In addition, the components must also be thoroughly degreased prior to treatment. This is to prevent arc discharges during the heating up process.

It is apparent that the advantages far outweigh its disadvantages. Low maintenance and operation costs, coupled with the elimination of post-treatment processes and rejects, renders the process to be economically viable. It is envisaged that the process will soon replace most of the conventional methods of nitriding.

10.5 FUTURE TRENDS AND THE SCOPE OF PLASMA HEAT TREATMENT

Plasma nitriding is fast gaining recognition and industrial acceptance. Research in this area is ongoing and much has been achieved over the past decade or so. Improvement in the industrial application of the process includes features such as enhanced control by microprocessors, data collation and enhanced control of the heating up and cooling down rates. Efforts were also made by manufacturers to integrate the process into the overall production system.

With the introduction of carboneous gas with standard N_2-H_2 gas mixtures, plasma nitrocarburizing can be carried out effectively. Conventional hardening processes for powder metallurgy (PM) parts tend to produce unacceptable distortion. Rembges [9] tried plasma nitrocarburizing on automotive PM parts. Treated components were observed to have 4–12 microns of compound layer with minimum porosity. The distortion problem was solved and the parts exhibited significant improvement in hardnesses, wear, and fatigue resistance.

BCl_3-H_2-Ar gas mixtures of various proportions were introduced into a modified ionitriding unit to plasma boronize ferrous materials [10,11] at 700–900°C. The microstructures obtained so far proved to be similar to those conventionally pack- or paste-boronized. Bolyce et al. then ventured on to examine the effect of laser melting on boronized steels. The boronizing of ferrous materials in plasma can be traced back to the early 1970s [12]. To this date, however, not much headway has been accomplished in this area.

The plasma nitriding of titanium and titanium alloys has drawn much interest. Titanium has a strong affinity for nitrogen, and it has been established that the golden TiN nitride takes only a few minutes to form [13]. Carried out at temperatures between 800°C and 1000°C in N_2-H_2 gas mixtures, plasma nitriding produces a compound layer of TiN and Ti_2N with a diffusion layer of nitrogen in solid solution [13,14]. This increases the surface hardnesses and improves upon the wear resistances.

Industrial interest in plasma carburizing and plasma carbonitriding has aroused much research and development activities in these areas. Specimens were treated in the austenitic temperature range of 850°C to 1050°C in various gas mixtures comprised of H_2, N_2 and CH_4, before rapid cooling by forced convection or oil-quench [15,16].

10.6 REFERENCES

1 Shohet, J. L. *The Plasma State*, Academic Press, 1971.

2 Dressler, S. "Plasma Assisted Surface Treatment," *SME Int'l. Conf. Proc.*, Michigan, pp. MS89–446/1-MS89–446/20, 1989.

3 Edenhofer, B. "Physical and Metallurgical Aspects of Ionitriding," *Heat Treat. Met.*, 1:23–28, 1974.

4 Jones, C. K. and Martin, S. W., *Metal Progress*, 85:94–98, 1964.

5 Loh, N. L. "Plasma Nitriding and the Fatigue Properties of 722M27 Steel," Ph.D. Thesis, Liverpool, 1980.

6 Loh, N. L. "Plasma Nitriding of Steels," *SEAISI Quarterly*, 17–3:56–65, 1988.

7 Hombeck, F. "Forward View of Ion Nitriding Applications," *Proc. 1986 Int'l. Conf. on Ion Nitriding*, ASM.

8 Stefanides, E. J. "Ion Nitriding Extends Mold's Wear Life," *Design News*, 45(7):92–93, 1989.

9 Rembges, W. "Plasma Nitriding of PM Parts," *Metal Powder Report*, 43(11), 1988.

10 Bloyce, A., Dearnley, P. A. and Bell, T. "Boride Surface Modifications," *1st Int'l. Conf. on Surface Engineering, Vol. III*, Brighton, pp. 93–108, 1985.

11 Wierzchon, T., Pokrasen, S. and Karpinski, T., *Hart. Tech. Mitt.*, 38(2):57, 1983.

12 Gifford, F. E., "Plasma Silicon Nitriding and Iron Boriding," *Journal of Vacuum Science Technology*, 10(1):85–88, 1973.

13 Badini, C., Mazza, D. and Bacci, T. "Texture of Surface Layers Obtained by Ion Nitriding of Titanium Alloys," *Materials Chemistry and Physics*, 20:559–569, 1988.

14 Matsumoto, O., Komuna, M. and Kanzaki, Y., "Nitriding of Titanium in an R.F. Discharge II: Effect of the Addition of Hydrogen to Nitrogen on Nitriding," *Journal of the Less-Common Metals*, 84:157–163, 1982.

15 Rie, K-T, Lampe, T. and Eisenberg, S. "Plasma Carburizing and Plasma Carbonitriding—An Austenitic Thermochemical Surface Treatment," *1st Int'l Conf. on Surface Engineering, Vol III*, Brighton, pp. 121–132, 1985.

16 Lees, M. I. and Taylor, B. J., "Plasma Carburizing," Electricity Council Research Centre, Chester, NTIS, 1986.

CHAPTER 11

A Process Overview of
Laser Hardfacing

A. F. A. HOADLEY[1]
A. FRENK[1]
C. F. MARSDEN[1]

11.1 INTRODUCTION

THE selection of a material for a hostile environment is often a compromise between the surface properties, i.e., its resistance to corrosion or wear, and the mechanical properties of the bulk material. Therefore, the possibility of modifying the surface properties by alloying or cladding with a different material has been of industrial interest for many years. Several processing techniques are available, such as MIG and TIG welding plasma-transferred arc, plasma spraying, combustion spraying, and laser surfacing. Although all of the above are thermal coating processes, they differ significantly in the bonding at the interface between the surfacing compound and the substrate. Three possibilities exist: mechanical bonding through plastic deformation, diffusion bonding between the liquid cladding material and the solid substrate, and a fusion bond, which alloys the clad to the substrate. Of these, only the latter guarantees a low level of porosity and strong adhesivity, both of which are desirable for hardfacing applications. Mixing of the substrate with the cladding material leads to a change of the clad composition and potentially a degradation of the clad properites. Due to the precise control over the laser energy, laser treatment is arguably the only means of producing a fusion-bonded clad in a single layer with minimal dilution. Although this gives the laser a clear advantage over other surfacing technologies, defects such as porosity, cracking, and residual stress formation are of equal importance in evaluating the process, together with the surface quality and processing time.

[1]Laboratoire de Métallurgie Physique, Ecole Polytechnique Fédérale de Lausanne, MX-G, Ch-1015 Lausanne, Switzerland.

The aim of this chapter is to present the process parameters, which are chosen in order to optimise laser cladding. In attempting to write an overview, it is necessary to acknowledge the pioneering work of Weerasinghe and Steen in the development of the laser cladding process [1]. They were also the first to model the blown powder method [2], and these ideas ultimately led to a control strategy for the process [3]. This chapter will address the subject from a different angle. It will consider "cause and effect" in such a way as to uncouple the influence of the parameters as far as possible. Where possible, experimental results will be presented, but in addition, computer calculations will also be used to demonstrate, in an unambiguous way, the influence of a specific parameter, which is not possible to isolate in an experimental study.

Finally, experience is drawn from using a 1.5 kW CO_2 laser, and although other types of lasers and laser systems are similar, there exist differences in the energy absorption and the intensity threshold for the onset of plasma. Efforts for hardfacing have been concentrated on the Stellite alloys. These cobalt-based compounds are known for their excellent strength, hardness, and corrosion resistance, particularly at high temperatures [4].

11.2 PROCESS DESCRIPTION

There are two fundamentally different methods for cladding by laser: the preplaced powder method and the blown powder method. The former is a two-step method, where the powder is first predeposited, either with or without a binding agent, prior to heating with the laser. Due to the poor thermal properties of the porous layer, the degree of dilution is difficult to control, and gas inclusions are often a problem. In the blown powder method, the clad powder is injected directly into the molten pool produced by the laser. This ensures intimate mixing of the powder with the liquid and therefore rapid heat transfer and fusion. The blown powder process is depicted schematically in Figure 11.1. This chapter will deal exclusively with this method, which is the superior process for all applications. Three main process components are involved: a laser system including a means of focussing, CNC-table, and a powder delivery system. In the following sections, each component will be described in terms of its function and the processing options that must be specified in configuring the system as a whole.

11.2.1 THE LASER SYSTEM

CO_2 laser radiation may be considered as a surface heat source due to its very low penetration depths in metals. If the total power output is fixed,

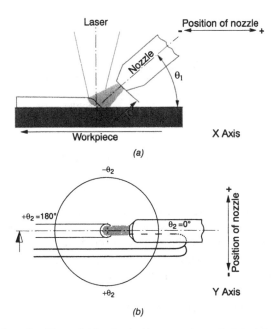

Figure 11.1 Schematic of laser cladding by the blown powder method depicting the workpiece and the powder injection nozzle: (a) a longitudinal section showing the Θ_1 inclination and the relative position of the nozzle; (b) a plan view of the clad surface showing the Θ_2 orientation and position in the y axis.

ideally at its maximum rating, then the intensity of the laser source is manipulated by altering the shape and area of the beam. The most common focussing systems produce circular, linear, or rectangular beams. A circular beam has an intensity that is a function of the radial distance from its centre. By contrast, a linear beam has a uniform intensity over a given width in the transverse of y direction but may be focussed in the longitudinal or x direction. The overall beam shape is controlled by the optics of the focussing mirror or lens. The beam area or spot size is altered by varying the distance between this focussing device and the component surface. Although the same energy intensity may be achieved before or after the focal point, operating below may produce plasma at the point of passing through the focal plane.

11.2.2 THE CNC TABLE

To cover a given surface, it is necessary to move the component relative to the laser beam. Two possibilities exist: either the specimen is fixed and the laser head moves, in which case the laser optics have to adjust for this displacement; or, the specimen is clamped to a numerically controlled

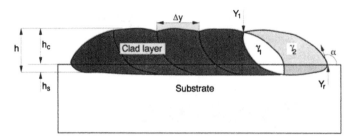

Figure 11.2 Schematic of a transverse section through a laser clad showing the overlapping tacks.

table, which moves relative to a fixed laser. Unless the covered region is very narrow, it is normally necessary to overlap a series of tracks, as illustrated in Figure 11.2. Hence, the surface coverage rate is controlled by two parameters set by the operator: the relative velocity between the laser and the component, v_x, and the inter-track advance, Δy. In the case where a cylindrical surface is cladded, an additional rotating axis is required. The previous parameters now correspond to the rotational velocity and the axial displacement.

11.2.3 THE POWDER FEEDING SYSTEM

Whereas the CNC-table is essential for most automated laser applications, the powder feeding system is specific to laser cladding and alloying. It must be capable of delivering a precise quantity of powder to the interaction zone produced by the laser. The powder delivery may be gravity fed or be a gas pressurised system. The latter is preferred since the higher particle velocities are not strongly influenced by gravity, thus allowing cladding in any orientation, while at the same time, the carrier gas has a protective effect inhibiting oxidation. The injection nozzle is clearly an important component of the powder feeding system. The aim of the nozzle is to direct the powder to the molten pool with the least possible dispersion. Hence, its design is dependent on the mass flow rates of both the powder and the gas. Figure 11.1 illustrates the three axes chosen by operator for the alignment of the nozzle relative to the point of impingement of the centre of the laser beam on the substrate surface.

11.3 PROCESS PERFORMANCE

In the preceding section, the process was described in terms of the three major components. With each of these was associated a number of adjustable parameters, which are set by the system operator. In addition, the maximum laser power is fixed by the installation, and normally, a minimum clad thickness, h_c, is specified by the customer. With these constraints and assuming that cladding of the two materials is thermodynamically feasible, then the parameters above may be selected to give a processing window. The aim of the following sections is to discuss this window within the framework of maximising the process performance.

11.3.1 THE POWDER EFFICIENCY

The powder efficiency is the ratio of the mass of powder participating in the clad layer to the mass of powder injected. It is important for two reasons. Firstly, powder lost from the process may be difficult to recycle, and therefore there is an enonomic incentive to maximise the yield. Secondly, it will be shown in Subsection 11.3.2 that the required processing power is a function of the deposition rate, where the deposition rate is simply the powder efficiency multiplied by the powder feed rate. Thus, control of the process will be facilitated when the powder efficiency does not vary strongly on changing the other processing parameters.

Only the powder particles impinging on the melt surface are absorbed into the clad; those hitting a solid surface are reflected and lost from the process. Therefore, the powder yield is controlled primarily by two variables, the trajectory and dispersion of the powder jet and the projected area of the melt pool in the plane normal to the axis of the jet. In turn, the powder dispersion is determined by the design of the injection nozzle and the diameter of its outlet and the mass flow rate of the powder and the gas. Clearly, when the nozzle size is larger than the molten area, then the powder efficiency will always be much less than unity. Assuming that the liquid surface area is essentially controlled by the energy distribution of the laser beam, then for a circular beam, Figure 11.3 is a plot of the measured powder efficiency for cladding Stellite 6 using a 2 mm injection nozzle. It is evident from these results that the efficiency is improved as the area of the laser beam, and therefore the area of the molten surface increases relative to that of the powder jet.

When the powder jet is larger than the melt area, the alignment of the nozzle can be critical for obtaining the maximum powder efficiency. An experimental study was conducted to determine the optimum injection angle, Θ_1 and Θ_2, for a circular nozzle and a circularly focussed laser beam. The results in Figure 11.4 show that increasing the injection angle in the vertical sense led to an increased efficiency, while in the horizontal plane, the best results were achieved with the powder injected from a position in front of the current track [5]. Both these results lead to the conclusion that the highest efficiencies are obtained when the nozzle is aligned to max-

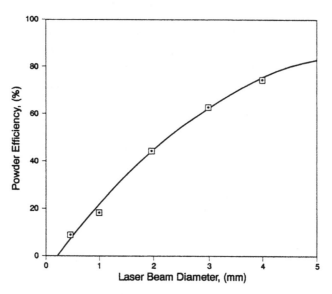

Figure 11.3 The influence of the beam diameter (circular focus) on the powder efficiency.

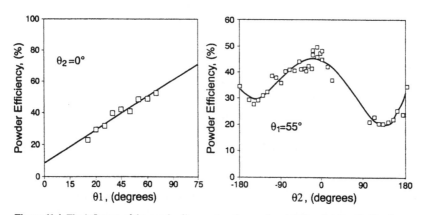

Figure 11.4 The influence of the nozzle alignment on the powder yield for cladding Stellite 6 onto mild steel with a 2 mm nozzle and a 2 mm laser beam diameter.

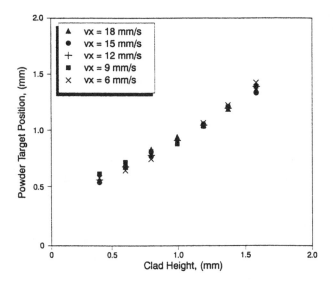

Figure 11.5 The calculated target position for the centre of the powder jet relative to the centre of the laser beam, for Stellite 6, injection angle 45°.

imise the projected surface of the molten pool. Considering now the position of the nozzle, a target position on the substrate must be found relative to the centre of the laser distribution. From this point, it is possible to locate the nozzle for a given alignment and a fixed injection distance. Finding the optimum target position is not so straightforward. If one assumes that a line between the nozzle and this target position on the substrate should intersect with the centre of the projected molten surface, then it is clear that this position will depend on the clad height and also on the position of the laser relative to the clad. This is exactly what is indicated in Figure 11.5, which is obtained from computer simulations of the melt surface under different processing conditions [6].

11.3.2 SURFACE COVERAGE AND CLAD DEPOSITION RATE

The processing time is the time to deposit a given thickness of clad onto the component surface. Clearly, the shorter the processing time, the greater the throughput from the laser installation and thus the maximum return on the invested capital. When the clad thickness is specified, then the surface coverage rate (the product of v_x and Δy) and the deposition rate are synonymous. However, it is possible to clad several layers in order to achieve the desired clad thickness. It is this strategy that will be discussed first.

Figure 11.6 presents the results from a thermal model of laser cladding described in detail in Reference [6]. In this two-dimensional finite element calculation, the clad height was specified and the laser power was given by a Gaussian distribution. Furthermore, the absorbed power was adjusted in order to produce a clad with almost no melting in the substrate. Figure 11.6 is therefore a map of processing conditions, which relate the required power to the clad height and the processing speed. Now considering an available power per unit length of 300 kW/m, then operating at point A would produce a clad of 1 mm in a single pass with a deposition rate per unit length of 6 mm²/s. However, it is equally possible to produce this total thickness operating at point B or C, although this leads to an effective rate of only 5 mm²/s and 4 mm²/s respectively. Therefore, it is possible to conclude that the greatest coverage rate is achieved when the clad is deposited in a single layer.

As it is always necessary to overlap adjacent clad tracks to form a continuous coating it is worthwhile to consider the influence of this parameter of the processing time. Defining the overlap, R_y, as

$$R_y = \frac{(S_y - \Delta y)}{S_y} \text{ where } 0 < R_y < 1$$

where S_y is the width of the track. In Figure 11.2, this is the horizontal distance between the points y_l and y_r. The total shaded area in Figure 11.2

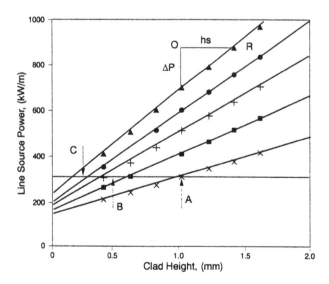

Figure 11.6 The calculated power required for cladding Stellite 6 onto a substrate of the same material ($D < 0.5\%$), the symbols correspond to Figure 11.5 laser absorption 20%.

represents the cross section of the liquid pool (γ_1 and γ_2), but the region γ_1 was deposited during the previous pass and is therefore remelted. Increasing the overlap increases the ratio of γ_1 to γ_2 and, therefore, decreases the amount of energy available for depositing new material. Therefore, it is possible to conclude that the deposition rate is adversely affected by increasing the overlap. However, as increasing the overlap always increases the height of the clad, the reduction in the deposition rate due to Δy may be partially offset by an increase in the processing speed V_x, in order to maintain a constant clad thickness. This may be illustrated in Figure 11.6, if one takes the height clad in each pass to be approximately $(1 - Ry)xh_c$. Ultimately, however, the overlap is chosen by the desired surface roughness.

11.3.3 SURFACE ROUGHNESS

The roughness of the clad surface is made up of two components. The macro roughness, which is a result of overlapping adjacent tracks producing a peak and a depression, gives rise to a periodic variation in the clad height (see Figure 11.2). The micro roughness, which is on a scale much smaller than the track width, is related to powder phenomena. Where the surface roughness exceeds the client's specification, then the surplus material must be removed in a machining or grinding operation. Not only is this wasteful of the hardfacing material, in the case of very hard clad layers, this additional operation may be very expensive, requiring diamond grinding. Hence, there is an incentive to produce a smooth surface, avoiding an additional process step where possible, or at least keeping to a minimum the amount of clad material that must be removed.

With metallic melts, surface tension forces dominate and therefore the free surface in the transverse direction will be close to a segment of a circle. Assuming that the width of the molten surface, S_y, is given by the laser beam diameter, then the maximum variation in height across each track may be calculated from a simple geometrical model. Figure 11.7 shows the calculated inter-track roughness as a function of the overlap for a clad height of 1 mm. It is clearly demonstrated that the macro roughness is primarily controlled by the overlap. As the overlap increases, the macro roughness is reduced, and when the overlap is greater than 80%, the roughness is essentially independent of the beam width.

The micro roughness has an important influence on the surface finish of the clad component. As the melt pool surface is essentially smooth, the cause of asperities is perhaps not evident. It is necessary to consider what occurs to the powder particles missing the melt pool surface. When the powder is injected from the front of the melt as indicated in Figure 11.1(a) ($\Theta_2 = 0°$), then those particles having too steep an injection angle will

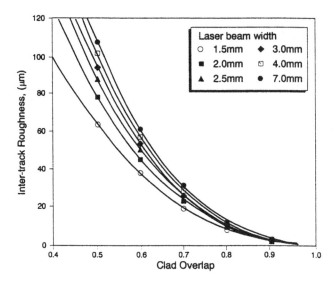

Figure 11.7 The theoretical inter-track roughness based on a circular surface geometry and a fixed clad height of 1 mm.

completely miss the laser beam and will bounce away on impact with the cold substrate in front of the clad. However, the particles with a shallow injection angle will pass through the laser beam before impacting with the surface of the clad behind the melt pool. The latter situation leads to perfect conditions for mechanical bonding on impact, as the surface immediately behind the molten clad is still very hot and the particles arrive preheated from the laser. Therefore, this injection orientation leads to a high micro roughness and a relatively poor surface finish. By contrast, when the powder is injected from behind the molten pool ($\Theta_2 = 180°$), hot powder, which has passed through the laser beam, impacts with the cold substrate surface and only cold powder lands on the hot rear surface of the clad, neither condition leading to mechanical bonding. The micro roughness and overall surface appearance for this orientation is therefore much improved by comparison to the previous orientation.

11.4 THE CLAD INTEGRITY

The aim is to produce a clad of high integrity, such that the desired properties of the clad material are in no way degraded by the process. In this section the influence of the processing parameters on the metallurgical quality of the clad layer is discussed.

11.4.1 INTER-TRACK POROSITY

Porosity occurs in laser cladding due to gas absorbed on the powder and oxidation of the powder, both of which can result in gaseous products trapped in the clad. Porosity also occurs due to poor wetting between the substrate and the clad. Whereas the former produces spherical pores in the clad layer and is related to the type of powder, the latter form of porosity manifests itself at the intersection between adjacent tracks and the substrate and results from the processing conditions. It is this inter-track porosity that will be discussed here.

Inter-track porosity occurs when the angle between the clad liquid surface and the substrate is steep (α in Figure 11.2) and normally is only found when the melt depth in the substrate is very low (on the limit of having a fusion bond at the interface). Porosity forms between tracks because this is where the interface angle is steepest, making wetting the most difficult, while the energy incident on the projected area of this surface normal to the beam is insufficient to remelt both the previous track and the substrate. Therefore, the inter-track angle may be used to gauge the likelihood of porosity formation. Assuming once again that the clad width, S_y, is given by the laser beam diameter, then Figure 11.8 is a plot of the inter-track angle for a clad thickness of 1 mm. It is seen here that the inter-track porosity is controlled by the width, whilst the overlap has only a minor effect.

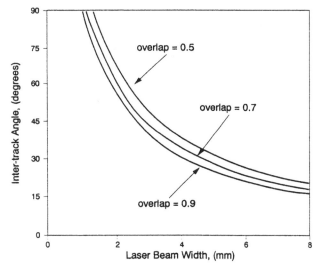

Figure 11.8 The theoretical inter-track angle based on a circular surface geometry and a fixed clad height of 1 mm.

11.4.2 THE CLAD DILUTION

The clad dilution is the primary control on the quality of a defect-free layer. It is the quantity of the substrate material that has been melted and has mixed with the clad and therefore may be calculated from the melt depth in the substrate h_s (see Figure 11.2),

$$D = \left(\frac{h_s}{h_s + h_c} \right) \times 100\%$$

Figure 11.9 shows the effect of iron on the hardness of a Stellite coating. As the iron is not present in the Stellite powder to any significant degree, the existence of iron in the clad is due to mixing with the substrate. Up to a composition of 5% wt of iron, there is little influence, the microhardness remaining close to 650 Hv; however exceeding this limit shows a marked effect, with the microhardness falling to about 500 Hv at only 15% wt. At the same time, it is important to recognise that some dilution is always necessary in order to ensure that a metallurgical bond exists between the clad and the substrate.

Melting into the substrate may be viewed in two ways, either that too much energy is incident on the clad, or too little powder was absorbed. In

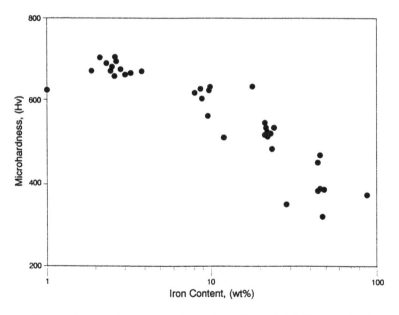

Figure 11.9 The influence of iron on the microhardness of laser clad Stellite 6, applied load 0.2 kg.

considering the surplus energy, it is necessary to return to Figure 11.6, which presents the idealised case where the power was adjusted to give minimum ($D < 0.5\%$) dilution. The intercept point for each line, clad height zero, is the power required to heat the substrate to the melting temperature. Increasing the power above this value allows cladding to proceed. The optimum power, at any given processing speed, increases linearly with the clad height. Thus, operating above this line will inevitably lead to increased dilution. Thus, if the operation point is point "O" in Figure 11.6, then ΔP is the vertical distance above the zero dilution line. However, if the power is fixed, then the true operating point is at the horizontal intersection, point R in Figure 11.6. This Δh corresponds to a deficiency in powder and, to a first approximation, may be related to the melt depth, h_s, allowing an estimate of the dilution. Increasing the powder feed rate should push the operating point to the right reducing the dilution. However, this leads also to a thicker clad, and therefore to return to the desired clad height, it is also necessary to increase the processing speed.

Finally, the dilution may also be controlled by laying down more than one layer. For example, where the dilution of the first layer is 10%, then depositing a second layer of the same thickness will lead to a composition on the surface containing only 1% of the substrate material. However, with this strategy, as mentioned earlier, the processing time is adversely affected.

11.4.3 RESIDUAL STRESSES, PRE- AND POST-HEAT TREATMENT

Perhaps the most important problem facing laser cladding is the production of stored stresses. This is particularly the case when hard material is deposited due to its low ductility. These stresses occur due to high thermal gradients and cooling rates during processing, which can result in cracking and deformation of the component and the formation of residual stresses on cooling to ambient temperature. The processes leading to the formation of these stresses are extremely complex. However, qualitative understanding can be obtained, assuming that the cladded workpiece is made up of rigidly joined layers of distinct materials, each with its own thermophysical properties. These successive layers compensate for the deformation produced during cooling (and heating) by formation of stresses. With laser treatments, the component is only superficially heated, and therefore the bulk of the substrate remains relatively cold, effectively blocking the deformation. On cooling, the cladded layer builds tensile stresses that are directly related to the thermal expansion coefficient of the cladded material and the temperature difference between the melting point of the clad and the temperature of the substrate, the processing temperature. Clearly, plastic deformation and creep will have a countering influence by partially relaxing these stresses. The second step in the residual stress formation occurs during cooling from the processing temperature down to the ambient temperature. In this case, both the expansion coefficients of the cladded material, as well as that of the substrate, have to be considered. When cladding a material with an expansion coefficient lower than that of the substrate, the tensile residual stresses produced in the clad may be converted into compressive stresses, depending on the processing temperature, which can be varied using preheating. This is the case, for example, when cladding WC-Co layers onto steel substrates.

The possibility of modifying the residual stress state in the clad layer using a post–heat treatment depends largely upon the difference in the thermal coefficients of expansion of the substrate and surfacing alloy. When the coefficient of expansion of the clad is smaller than that of the substrate, as is the case of Stellite 6 ($\alpha = 16 \times 10^{-6}$ K^{-1}) on stainless steel ($\alpha = 19 \times 10^{-6}$ K^{-1}), a post–heat treatment can convert the tensile stresses generated in the clad into compressive ones, provided a high enough post–heat treatment temperature can be achieved during which all the stresses are relieved through plastic deformation and creep. Therefore, on cooling again to room temperature, the higher thermal expansion of the substrate material produces compressive stresses in the layer. This was shown by the results of Dekumbis [7] for a cylindrical coating, where the geometry precludes stress relief by deformation (see Figure 11.10). In contrast, stress relieving by a post–heat treatment of the stresses generated when cladding a mild steel ($\alpha = 14 \times 10^{-6}$ K^{-1}) with Stellite 6 can only

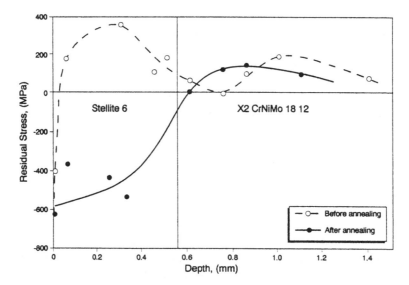

Figure 11.10 Residual stresses measured by X-ray method in laser cladded Stellite 6 on steel X2CrNiMo 18 12 before and after annealing at 900°C.

occur by deformation of the cladded component (i.e., bending of a plate). In the case of a cylinder, no inversion of the sign of the stress field is possible on post–heat treatment.

11.5 CONCLUSIONS

The objective of this overview was to show that, in laser cladding, there is no single operation point that leads both to peak process performance and the highest level of clad integrity. Instead, an operation window exists that allows the process to be optimised, by weighing a number of criteria. However, it is still possible to draw a number of conclusions:

- The maximum powder efficiency is achieved by maximising the projected area of the liquid surface with respect to the powder jet diameter.
- There is little point in producing a clad with low macro roughness using a high level of overlap if the micro roughness determined by the particle size and injection geometry has a greater influence.
- Inter-track porosity occurs when the inter-track angle is too steep, and therefore it may be controlled by increasing the laser beam width and reducing the cladding speed as a consequence.
- Dilution occurs when insufficient powder has been absorbed, and hence it may be reduced by increasing the powder feed rate. When it is only the surface composition that is important, the dilution may also be improved by depositing more than one layer.
- Under some circumstances post–heat treatment provides a means of modifying the residual stress state and thereby turning detrimental tensile stresses in the clad to compressive stresses. Preheating may also be used to reduce the stored stresses produced by the process.

11.6 ACKNOWLEDGEMENTS

The authors would like to thank the Swiss government, "Commission pour l'Encouragement de la Recherche Scientifique," Berne and Sulzer Innotec, Winterthur, for financial support with this work.

11.7 REFERENCES

1 Weerasinghe, V. M. and Steen, W. M. "Laser Cladding with Pneumatic Powder Delivery," *Lasers in Materials Processing,* E. A. Metzbower, ed., American Society of Metals, Ohio, pp. 166–174, 1983.

2 Weerasinghe, V. M. "Computer Simulation Model for Laser Cladding," *Transport Phenomena in Materials Processing,* M. M. Chen, ed., ASME, New York, pp. 15–23, 1983.

3 Li, L. and Steen, W. M. and Hibbert, R. D. "Computer Aided Laser Cladding," *3rd European Conference on Laser Treatment of Materials ECLAT 90,* H. W. Bergmann and R. Kupfer, eds., AWT-Arbeitsgemeinschaft, pp. 535–542, 1990.

4 Antony, K. C. "Wear-Resistant Cobalt-Based Alloys," *J. of Metals,* pp. 52–60, 1983.

5 Marsden, C. F., Frenk, A., Warnière, J.-D. and Dekumbis, R. *3rd European Conference on Laser Treatment of Materials ECLAT 90,* H. W. Bergmann and R. Kupfer, eds., AWT-Arbeitsgemeinschaft, pp. 355–369, 1990.

6 Hoadley, A. F. A., Marsden, C. and Rappaz, M. "A Thermal Model of the Laser Cladding Process," *22nd ICHMT on Manufacturing and Materials Processing,* Dubrovnik, Hemisphere, New York, 1991.

7 Dekumbis, R. "Controlling Residual Stresses in Laser Cladded Coatings," *6th International Conference on Lasers and Manufacturing,* Birmingham, 1989.

CHAPTER 12

Plasma Immersion Ion Implantation at Elevated Temperatures

R. HUTCHINGS[1]
M. J. KENNY[2]
D. R. MILLER[3]
W. Y. YEUNG[4]

12.1 INTRODUCTION

THE process of ion implantation provides a versatile and controllable method for modifying the surface composition and properties of materials. Although it has been shown to be very effective in improving the wear resistance of a range of metals [1,2], its widespread application has been hindered by both real and perceived limitations [3]. Among these are the extremely shallow treatment depth (of the order of 100 nm) and the need for the target and/or the beam to be manipulated to achieve uniform surface coverage. The technology required for a reliable high current ion source has also meant that implantation facilities have generally been restricted to major research laboratories.

Recently, a technique has been developed that offers the possibility of taking ion implantation out of the specialised laboratory and onto the toolroom floor. The target is placed in a plasma and biased to a high negative voltage. Ions from the plasma are accelerated towards the target and implanted approximately uniformly over the entire exposed surface, eliminating the need for target manipulation or beam rastering. This technique has several distinct advantages over traditional methods of implantation, including uniform coverage, low unit cost, the easing of line-of-sight re-

[1]Australian Nuclear Science and Technology Organisation, Private Mail Bag 1, Menai, NSW 2234, Australia.
[2]CSIRO, Division of Applied Physics, Private Mail Bag 7, Menai, NSW 2234, Australia.
[3]Department of Chemical Engineering, University of Adelaide, G.P.O. Box 498, Adelaide, SA 5001, Australia.
[4]NM Metals, P.O. Box 21, Port Kembla, NSW 2505, Australia.

strictions, and the ability to treat complex shapes. In addition, it has the ability to scale to large targets and offers the prospect of a technologically simple implanter design. The technique was first proposed by Conrad and his coworkers at the University of Wisconsin, using a filament discharge plasma [4], and independently developed at the Australian Nuclear Science and Technology Organisation (ANSTO) using an inductively coupled radio frequency (rf) glow plasma [5]. We believe that there are some inherent advantages in the rf version of the technique, which we call Plasma Immersion Ion Implantation (PI^3).

The first limitation mentioned above, the shallow treatment depth, is fundamental to the process of ion implantation. High-energy ions (typically in the range 20–200 keV) are fired into the surface of the component being treated and lose their energy through collisions with atoms in the crystal lattice. They come to rest at positions forming a roughly Gaussian distribution of depths below the surface. The peak of the ion distribution will be at a depth depending on the energy and mass of the implanted ion and the composition and density of the substrate. For 50 keV nitrogen ions implanted into steel, the ion distribution will have a peak at a depth of ~ 60 nm and a half-width ~ 100 nm (FWHM). Typically, ion implantation is carried out at low temperatures (below 200°C), and room temperature processing is possible.

Although extremely shallow, the treated layer can possess unique properties. Since ion implantation is a nonequilibrium process, very high concentrations (more than 50 atom%) of the implanted ion can be obtained, and it is possible to incorporate ions with a gross misfit into the crystal lattice. Considerable damage is introduced due to displacement of the substrate atoms, and states of intermediate crystallinity or even amorphous zones can be produced. The incorporation of the implanted ions also results in a region of high compressive stress close to the surface. Despite the shallowness of the treated layer, changes in the macroscopic properties of the material are possible. Nitrogen implantation can increase the surface hardness by interstitial hardening or by nitride precipitation. This increased hardness can improve the wear resistance, but dramatic increases in wear resistance are obtained when implantation triggers a transition from the normal mode of wear to a less severe mode [3].

However, it must be admitted that the production of a thin surface layer with superior tribological properties will not always be enough. The underlying material must have sufficient load-bearing capacity. This has motivated the investigation of implantation at higher temperatures where some diffusion of the nitrogen to depths greater than 100 nm might be expected. Until recently, most of these investigations failed to produce such a result [6]. For temperatures in the range 50–200°C, the nitrogen was observed to migrate progressively towards the surface while keeping the total nitrogen content constant. Even at high doses implanted by conventional ion beams at 200°C [7], when the nitrogen content saturates, nitrogen is still confined to its expected implantation depth. At temperatures above 200°C, the nitrogen content decreases, corresponding to diffusion out of the implanted region. Similar results have also been obtained by annealing after implantation [8].

Early PI³ results [5], however, revealed some cases in which nitrogen penetrated to depths much greater than expected. Despite the relatively low implantation energies (20 keV N_2^+ ions), significant increases in wear resistance and surface hardness were obtained and exceeded those achieved by conventional ion implantation. Subsequent investigation [9] has shown that this is due to the diffusion of nitrogen to depths of greater than 1 μm at elevated temperatures ($T > 250$°C). The implanted layer is maintained and is supported by an underlying diffusion zone.

This chapter reports on the continuing investigations of implantation at elevated temperatures and discusses the effects produced in a range of steels, specifically 0.3% carbon mild steel, austenitic stainless steel, and 5% Cr tool steel. These results are compared with those of conventional ion beam implantation and present preliminary results where attempts have been made to mimic some of the conditions of PI³ treatment in a conventional implanter. The current results are compared with those obtained by plasma nitriding, even though the present treatment temperatures are still well below those generally used for plasma nitriding.

12.2 EXPERIMENTAL DETAILS

The present version of the PI³ system at ANSTO is based on an inductively coupled rf plasma produced in a cylindrical vacuum vessel (diameter 30 cm, length 40 cm) made of borosilicate glass. Approximately 250 W of rf power at 12–13 MHz is used to produce the plasma at a gas pressure of 10^{-3} mbar. Theoretical calculations indicate that the dominant ion species is N_2^+ with ~ 10% of N^+. The workpiece forms the cathode of the bias circuit, which supplies −45 kV pulses. One of the aluminium end-plates that seals the glass cylinder is grounded and forms the anode of the implantation circuit. A schematic of the apparatus is shown in Figure 12.1. The pulse repetition rate is varied to control the surface temperature of the sample. Details of the operation of the implanter are given in Reference [9].

Conventional implantation was carried out by the CSIRO on a 50 keV research implanter with a Freeman ion source and analyser magnet. Modifications to the implanter were made to imitate some of the conditions encountered in PI³. A heater stage enabled implantation to be carried out at

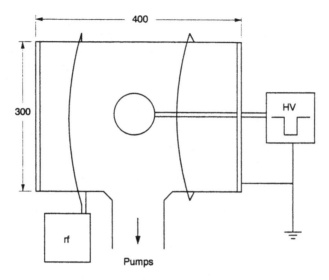

Figure 12.1 Schematic of the PI³ apparatus.

temperatures up to 300°C and the pressure in the target chamber was raised from its normal value of 1.5×10^{-4} mbar by partially backfilling with nitrogen.

Retained nitrogen profiles are measured by use of the $^{14}N(d,\alpha)^{12}C$ nuclear reaction. The structural changes in the surface layer are investigated by glancing angle X-ray diffraction. Surface hardness is measured using both Vickers and Knoop indenters at loads of 15, 50, and 200 g. Microhardness profiles are obtained from cross-sectional samples. Wear measurements are made on pin-on-disk machines using a fixed ball as the pin. Wear track depths are determined using either interference microscopy or profilometry.

12.3 MILD STEEL

Disks of diameter 25 mm and thickness 5 mm were cut from cold drawn, low carbon (0.3%), steel bar stock (AS1443-1983) and mechanically polished to a 1-μm diamond finish and supported freely in the PI³ chamber. Nine mild steel samples were implanted at $V = -25$ kV with pulse repetition rates of 20, 33, and 50 Hz (resulting in final temperatures of 200°C, 290°C, and 340°C) with three different total implantation times of $N\tau = 5$, 10, and 15 s. This corresponds to implantation doses of $D_{imp} = 2.2$, 4.4, and 6.6×10^{17} atoms $\cdot$ cm^{-2}.

12.3.1 RESULTS

Figure 12.2, shows the depth profiles obtained by nuclear reaction analysis for samples at the extremes of implantation dose and temperature. In each of the nine samples, there is a peak some 50 nm below the surface, but the poor depth resolution (approximately 60 nm at the surface) does not allow direct comparison of the peak location with the expected implantation depth. (For 25 keV N_2^+ ions, standard theories [10] predict a concentration profile peaking at ~15 nm with FWHM of ~25 nm.) The observable peak broadening should be dictated by the measurement resolution but only the lowest dose, lowest temperature sample, has a peak with FWHM of the expected 60 nm. When the temperature and implantation dose increase, the peak broadens to give FWHM of ~100 nm, appreciably greater than that expected from resolution limitations alone. More significantly, a tail appears in the profile, indicating diffusion of nitrogen to depths greater than 1 μm [Figure 12.2(b)].

Figure 12.3, shows separately the amounts of nitrogen contained in the surface peak and the diffusion-enhanced tail. The nitrogen content of the peak appears to saturate at ~2×10^{17} atoms·cm^{-2}, which is well above the level expected for 25 keV N_2^+ ions but is consistent with the enhanced broadening of the peak. The nitrogen content of the diffusion enhanced tail, however, continues to increase with implantation dose as long as the temperature is above a threshold of ~260°C. Implantation doses greater than those of Figure 12.3 have shown no sign of saturation, with the

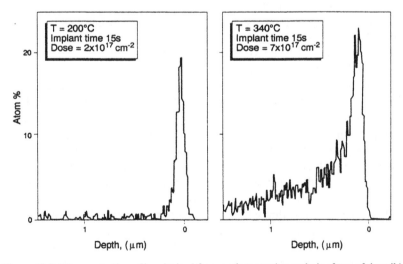

Figure 12.2 Nitrogen depth profiles obtained from nuclear reaction analysis of two of the mild steel samples.

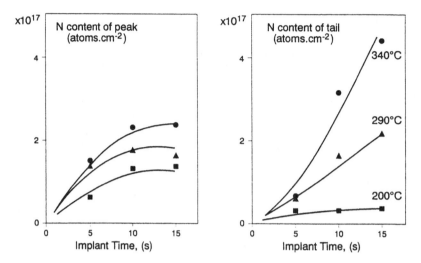

Figure 12.3 Nitrogen content of the implanted peak and diffusion enhanced tail as a function of implantation time and temperature in the mild steel samples.

diffusion-enhanced tail growing in height and depth as long as the temperature is above threshold.

Glancing angle X-ray diffraction has shown clear evidence of precipitates of γ'-Fe_2N_{1-x}. A typical diffraction pattern is shown in Figure 12.4 for the high dose, high temperature implant at an incidence angle of 1°. Varying the incidence angle from 0.5° shows that the ϵ-Fe_2N_{1-x} is more strongly concentrated at shallow depths.

All of the samples show increases in the surface hardness under a 15-g load. The unimplanted hardness was $H_v = 180$, while implanted specimens gave values of H_v up to 240. It should be borne in mind that this increase underestimates the hardening of the treated layer because of contributions from underlying, unmodified material. Figure 12.5 summarises the results of wear tests on some mild steel disks uning a tungsten carbide ball (10-mm diameter) and a 50-g load. Contact velocity was 0.8 m·s⁻¹ with paraffin oil continuously dropped onto the sample surface to provide light lubrication. An unimplanted specimen is compared with two specimens treated at high temperature. Both implanted specimens showed insignificant wear for the first hour, but after two hours the wear rate of the low dose sample ($D_{ret} = 1.4 \times 10^{17}$ atoms·cm⁻²) increased. This indicates that substantial breakthrough of the shallow modified layer had taken place. The high dose implant ($D_{ret} = 4.4 \times 10^{17}$ atoms·cm⁻²) continued to show a track with negligible wear even after four hours when breakthrough of the modified layer was evident. Wear testing was not continued since a flat had been worn on the tungsten carbide ball.

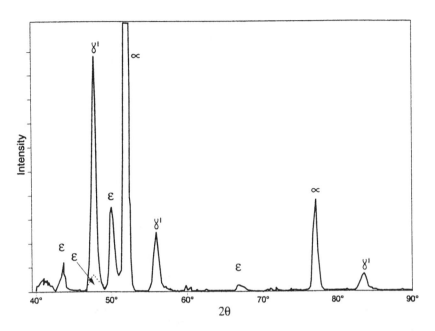

Figure 12.4 X-ray diffraction pattern obtained from a mild steel sample after high-dose, high-temperature implantation (incidence angle 1°).

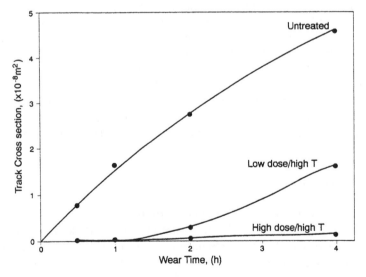

Figure 12.5 Wear characteristics of implanted mild steel discs.

193

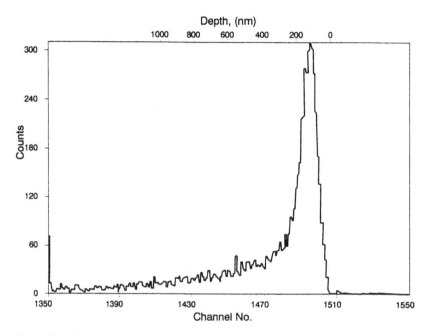

Figure 12.6 Nitrogen concentration profile obtained for a mild steel sample implanted in a conventional implanter with a backfill of nitrogen.

Preliminary results using conventional ion beam implantation are shown in Figure 12.6. Mild steel specimens were implanted with 25 keV ^{14}N ions to a nominal dose of 2×10^{17} atoms·cm^{-2}. Figure 12.6 shows the nitrogen concentration profile for a sample implanted with a backfill nitrogen pressure of 7.5×10^{-4} mbar. Whereas the expected range of ^{14}N ions in mild steel is 30 nm, this peak shows a substantial nitrogen concentration down to 300 nm and some nitrogen as far as 1 μm into the steel. The total retained nitrogen content D_{ret} is almost 4×10^{17} atoms·cm^{-2} or double the nominal dose. This is in complete contrast to the results obtained at high vacuum where almost no nitrogen is retained.

12.3.2 DISCUSSION

The most significant factor to emerge from the preceding results is the diffusion of nitrogen to depths of greater than 1 μm at elevated temperatures in contrast to the results obtained in conventional ion beam implantation under high vacuum. In PI^3, the total nitrogen content rises steadily with temperature and dose, corresponding to increased diffusion at depths beyond the normal implanted layer. This implanted layer is preserved even at temperatures of 340°C. We believe that the high activity of the nitrogen in the plasma plays a role in preventing the loss of nitrogen that would normally occur at high temperature. It has recently been reported [11] that similar retention of nitrogen can be accomplished by the use of ultra-high current density ion beam impantation. The mechanisms responsible for retention of nitrogen at elevated temperatures under such conditions are not clear and may well be different from those occurring during PI³ or conventional implantation with a backfill of nitrogen.

12.4 ALLOY STEELS

Small coupons 35 mm × 20 mm and thickness 1/3 mm were cut from a sheet of type 304 stainless steel and treated in the PI³. Disks of diameter 15 mm and thickness 3 mm were cut from heat-treated H13 tool steel (5% Cr content) and PI³ treated. Rods of diameter 2 mm were also prepared and implanted prior to sectioning.

12.4.1 RESULTS

Considerable nitrogen diffusion leading to large retained nitrogen doses was also observed in the H13 tool steel. Nominal dose and retained nitrogen concentration are given in Table 12.1 for samples implanted at $V = -40$ kV. At $T = 350°C$, there is significant diffusion, even at the lowest dose.

Glancing angle X-ray diffraction of implanted H13 revealed the formation of $\epsilon\text{-}Fe_2N_{1-x}$, although the 0002 reflection was absent in some cases. A least squares refinement of the hexagonal cell parameters for the data from sample 2 gives $a = 2.74$ Å and $c = 4.40$ Å. Microhardness measurements for implanted H13 are shown in Table 12.1 and indicate a steady increase of hardness with temperature and dose. The increase of hardness measured at 200 g is especially encouraging. Profiles of Vickers microhardness are shown in Figure 12.7 for sections cut from 2 mm rods of im-

TABLE 12.1. Retained Dose and Hardness as a Function of Temperature and Nominal Dose for H13 Tool Steel.

Sample	D_{imp} × 10¹⁷ cm⁻²	Treatment Temp (°C)	D_{ret} × 10¹⁷ cm⁻²	H_K (25 g)	H_K (200 g)
5	5	350	12.6	1240	696
1	8	350	13.9	1240	850
2	17	350	17.5	1550	1170
6	8	300	8.9	790	620
3	7	200	1.3	540	530

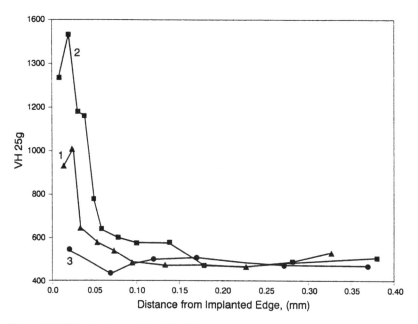

Figure 12.7 Microhardness profiles obtined from H13 specimens implanted under the conditions of 1, 2, and 3 in Table 12.1.

planted H13 tool. At the higher dose and temperatures, substantial hardening occurs down to 50 μm. Unfortunately, the nitrogen concentration at this depth could not be measured by the NRA technique used.

Excellent nitrogen retention was found for the type 304 stainless steel although indications are that the threshold temperature for the enhanced diffusion is somewhat higher than for mild steel. Figure 12.8 shows that the nitrogen concentration profiles for two samples implanted at $V = -28$ kV to a nominal dose of 15×10^{17} atoms·cm^{-2} at $T = 230°C$ and $340°C$. While the low temperature specimen shows only a surface peak with total retained nitrogen content of 2.5×10^{17} atoms·cm^{-2}, the amount of diffusion obtained at the higher temperature is remarkable, giving rise to a large retained nitrogen dose of 20×10^{17} atoms·cm^{-2}.

The diffraction patterns obtained from the type 304 stainless steel are shown in Figure 12.9. Here we compare two high temerature implants ($T = 340°C$), but at two different doses, $D = 6$ and 20×10^{17} atoms·cm^{-2}. A set of broad uncharacterised peaks appears to the left of each γ-Fe peak. These peaks broaden and shift towards larger lattice parameters as the dose increases. The broadness indicates considerable strain and the peak shifts are greater than those expected from an austenite-based (Fe,Ni,Cr)$_4$N in the high dose specimen. Significant im-

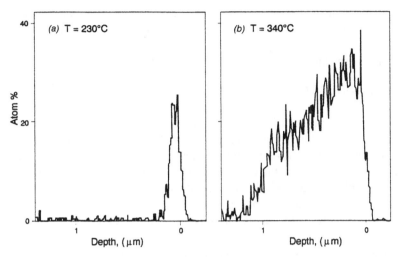

Figure 12.8 Nitrogen concentration profiles of type 304 stainless steel implanted to a nominal dose of 15×10^{17} atoms·cm⁻².

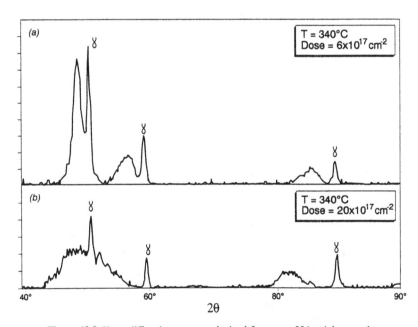

Figure 12.9 X-ray diffraction patterns obtained from type 304 stainless steel.

provements in microhardness were seen for the type 304 stainless steel. With a 15-g load, surface hardness was observed to increase from $H_v = 220$ for an unimplanted specimen up to $h_v = 310$ for the high-temperature, high-dose implant of Figure 12.8(b).

12.4.2 DISCUSSION

Again, the most striking result is the diffusion of nitrogen to depths greater than 1 μm at elevated temperatures. The concentration profiles obtained both in the austenitic stainless and the 5% Cr tool steel are much broader than those for the mild steel and have higher nitrogen content. In the case of the H13 steel, this may be attributable to the formation of fine CrN precipitates, which are too small to appear in the X-ray diffraction patterns, or simply to the high affinity of Cr for N even when present in solid solution.

The most obvious interpretation from the X-ray patterns for the type 304 stainless is that the austenite has been expanded by nitrogen in interstitial solution, but without precipitating equilibrium nitrides despite the high nitrogen content. This phase has been observed in ultra-high current density implantations mentioned above [11] but has also been observed in plasma nitriding of type 304 stainless steel at "low" temperatures (350–500°C). A full discussion is given in Reference [9]. This uncharacterised phase is responsible for causing a profound hardening of the surface and has also been ovserved in 310 stainless steel coatings deposited by reactive sputtering in Ar/N$_2$ [12]. The deposited layers can have very high nitrogen concentration (up to 50 atom%).

12.5 GENERAL DISCUSSION

It is of interest not only that the nitrogen diffuses to considerable depths during PI³ treatment, but also that the "retained" dose can exceed the nominal implanted dose. This suggests that adsorption of nitrogen and its subsequent inwards diffusion plays a role, leading to the comparison of PI³ with plasma nitriding. Even at the highest temperature, PI³ still lies at the low-temperature end of the plasma-nitriding regime. Below 500°C, nitriding rates for type 304 stainless steel are low, although nitride layer growth has been observed at temperatures down to 280°C [13] using a pulsed rf glow discharge. The gas pressures used in conventional plasma nitriding are usually greater than 1 mbar, at least three orders of magnitude higher than used in PI³. Recently, however, there has been some nitriding work using a PIG discharge plasma at low pressures comparable to those used in PI³ [14].

In comparing plasma nitriding to PI³, it should be noted that process times are generally a factor of 10 − longer than used in PI³. It seems that there are some inherent advantages of the low-pressure plasma for nitriding that should be studied further. These may be related to the reactivity and energy of the active species. It appears that PI³ should be regarded as a hybrid treatment − a surface layer with the superior tribological properties obtained by ion implantation supported by a thicker diffusion zone. PVD TiN coating on plasma-nitrided steel has been tried [15] in order to produce this type of effect, but PI³ has the advantage of accomplishing such a "duplex" layer in one processing step.

12.6 CONCLUSION

It has been shown that it is possible to obtain diffusion of implanted nitrogen to depths greater than 1μm, thereby producing significant improvements in surface hardness and wear resistance. Ion beam implantation by ultra-high current densities, or in a backfill of nitrogen gas, can give similar results but still suffers from line-of-sight and component manipulation problems. PI³ can lead to the widespread use of ion implantation as a viable surface modification technology in the same manner that plasmas have replaced ion beams in ion plating and ion-assisted deposition.

12.7 ACKNOWLEDGEMENTS

Support for this work is provided under the Generic Technology component of the Australian Industry Research and Development Act 1986. This chapter represents the efforts of a large number of people at the collaborating organizations, in particular John Tendys and George Collins (ANSTO), Leszek Wielunski and Ron Clissold (CSIRO), and Robert Finch (University of Adelaide).

12.8 REFERENCES

1 Dearnaley, G. "Ion Implantation and Ion Assisted Coatings for Wear Resistance in Metals," *Surf. Eng.*, 2:213–221, 1986.

2 Oliver, W. C., Hutchings, R. and Pethica, J. B. "The Wear Behaviour of Nitrogen Implanted Metals," *Metall. Trans.*, A15:2221–2229, 1984.

3 Legg, K. O. and Solinick-Legg, H. O. "Trends in Ion Implantation and Ion-Assisted Coatings," *Proc. 1st Int. Conf. on Surface Modification Technologies,*

Phoenix, Arizona, T. S. Sudarshan and D. G. Bhat, eds., TMS, Warrendale, Pennsylvania, pp. 79–96, 1988.

4 Conrad, J. R., Radtke, L., Dodd, R. A., Worzala, F. J. and Tran, N. C. "Plasma Source Ion-Implantation Technique for Surface Modification of Materials," *J. Appl. Phys.*, 62:4591–4596, 1987.

5 Tendys, J., Donnelly, I. J., Kenny, M. J. and Pollock, J. T. A. "Plasma Immersion Ion Implantation Using Plasmas Generated by Radio Frequency Techniques," *Appl. Phys. Lett.*, 53:2143–2145, 1988.

6 Moncoffre, N. "Nitrogen Implantation into Steels," *Mater. Sci. Eng.*, 90:99–109, 1987.

7 Terwagne, G., Piette, M., Bertrand, P. and Bodart, F. "Temperature and Dose Dependence of Nitrogen Implantation into Iron," *Mater. Sci. Eng.*, B2:195–201, 1989.

8 Barnavon, Th., Jaffrezic, H., Marest, G., Moncoffre, N. and Tousset, J. "Influence of Temperature on Nitrogen-Implanted Steels and Iron," *Mater. Sci. Eng.*, 69:531–537, 1985.

9 Collins, G. A., Hutchings, R. and Tendys, J. "Plasma Immersion Ion Implantation of Steels," *Mater. Sci. Eng.*, in press.

10 Ziegler, J. F., Biersack, J. P. and Littmark, U. *The Stopping and Range of Ions in Solids.* Pergamon, New York, 1985.

11 Williamson, D. L., Ozturk, O., Glick, S., Wei, R. and Wilbur, P. J. "Microstructure of Ultrahigh-Dose, Nitrogen-Implanted Iron and Stainless Steel," *Nucl. Instrum. Meth.*, in press.

12 Saker, A., Leroy, Ch., Foos, M., Michel, H. and Frantz, C. "Properties of Sputtered Stainless Steel–Nitrogen Coatings and Structural Analogy with Low Temperature Plasma Nitrided Layers of Austenitic Steels," *Mater. Sci. Eng.*, in press.

13 El-Hossary, F., Mohammed, F., Hendry, A., Fabian, D. J. and Szaszne-Csih, Z. "Plasma Nitriding of Stainless Steel Using Continuous and Pulsed rf Glow Discharge," *Surf. Eng.*, 4:150–154, 1988.

14 Nunogaki, M., Suezawa, H., Hayashi, K. and Miyazaki, K. "Plasma Source Nitriding," *Appl. Surf. Sci.*, 33/34:1135–1141, 1988.

15 Sun, Y. and Bell, T. "Plasma Surface Engineering of Low Alloy Steel," *Mater. Sci. Eng.*, in press.

PART II

CHARACTERISATION, PROPERTIES AND PERFORMANCE EVALUATION OF COATINGS AND SURFACE TREATMENTS

Observations on the Role and Selection of Coatings and Surface Treatments in Manufacturing

A. R. COOPER[1]
K. N. STRAFFORD[1]

13.1 INTRODUCTION

WEAR, in its various forms (refer to Tables 13.1 and 13.2), together with corrosion and fatigue, severely limits the useful life of engineering components. The cost of wear damage to the United States economy in 1980, for example, was estimated at a minimum of 20 billion United States dollars and possible as high as 100 billion United States dollars [1].

As wear is not an intrinsic material property, but characteristic of an engineering system, any change in load, speed, or environmental conditions can cause major changes in wear rate on one or both contacting surfaces [2,3]. Solutions depend upon the precise identification of the nature of the problem and great care must be exercised in applying general solutions to specific problems. The situation is further complicated by many inconsistencies and the lack of fundamental knowledge regarding the nature of the wide variety of wear mechanisms and the conditions under which they occur. To enable accurate predictions of wear and improve wear performance, there is evidently a need to

(1) More accurately define the wide variety of possible wear mechanisms and the particular conditions under which they occur [4,5]. Here the use and development of wear mechanism maps would appear to have particular merit [6].

(2) Establish a more applicable range of standard tests, evaluation methods, etc., to overcome the somewhat empirical and inconsistent

---

[1]Department of Metallurgy, Gartrell School of Mining, Metallurgy & Applied Geology, University of South Australia, Levels Campus, The Levels, SA 5095, Australia.

TABLE 13.1. **Various Wear Modes Encountered in Engineering and Manufacturing.**

Wear	Loss of material in form of fragments or debris and damage to component surfaces due to mechanical contract. Main interactions are loads and motions producing one or more of a number of abrasive, adhesive, or fatigue processes, all of which lead to wear debris.

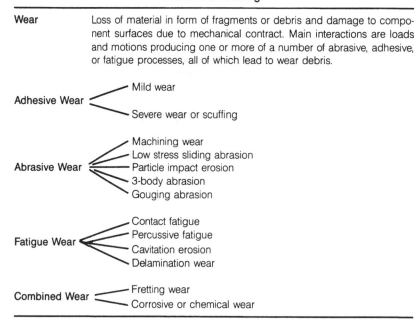

nature of wear tests and their sensitivity to test conditions. Here the wider, more systematic development of "round-robin" evaluations would appear desirable [7].

(3) Improve lubricant properties and the design of manufacturing tools.

(4) Research further into the role(s) of microgeometrical, macrogeometrical, tribochemical surface and metallurgical properties on wear processes [8].

(5) Provide guidelines for the selection and use of more wear resistant surfaces, be they bulk materials or advanced or improved coatings and surface treatments [9,10].

Additional advances could be made by investigating further the use of improved detection, quality control and reliability procedures, the nature of wear mechanisms in other systems (e.g., ceramics, polymers, composites), the types of wear debris, the size and distribution of real areas of contact between solids and finally the mathematical modelling and quantification of wear processes.

A study focussed on points (1), (2), and (5) is presently being conducted in the Department of Metallurgy at the University of South Australia [11]. This work, jointly supported by CSIRO and the University, is now being

TABLE 13.2. **Suggested Definitions for the Various
Wear Mechanisms Encountered.**

Adhesive Wear	Removal of material from a surface by (possible) welding together and shearing of minute areas or asperities when they slide (or roll) across each other under loads sufficient to cause local plastic deformation and adhesion. Some degree of sliding must take place and material may be transferred from one contact phase to another. Favoured by dry, chemically and physically clean surfaces, nonoxidising conditions, and chemical/structural similarities between the sliding couple.
Abrasive Wear	Removal of material from (and deformation of) a (relatively) soft surface when (relatively) hard material or asperities contact or impinge on it under load. The hard particles penetrate or indent the surface, thereby displacing materials by various mechanisms (cutting, ploughing, chipping, or fatigue cracking) depending upon material properties and the type of relative motion or loading. Particles may be attached to one of the surfaces (asperities or embedded) or may be loose, having been detached from one or both surfaces, e.g., by foreign particles in the lubricant.
Fatigue Wear	Wear process initiated and controlled by a fatigue mechanism and cyclic loads less than that normally required to permanently deform material. Fatigue wear leads to loss of metal particles from metallic surfaces and should not be confused with fretting wear, which involves production of primarily metal oxide particles.
Fretting Wear	Wear mechanism between tight-fitting contacting surfaces subjected to small, relative oscillatory motion or vibrational slip of extremely small amplitude. The process is characterised by production of very fine oxidised wear debris. Process can also be associated with conditions of vibratory impact in which percussive and fatigue wear may be present.
Chemical Wear	Mechanical wear damage in the presence of significant detrimental chemical reaction(s). For example, acid- or sulphur-containing fluids in contact with mechanical bearings (or similar moving contacting surfaces).

extended within a large GIRD project being carried out by the Department of Metallurgy [12]. The principal aim of the joint project is to try and critically cite, redefine, re-appraise, and analyse a range of wear problems and failures as noted in the open scientific and engineering literature. A number of the proposed hypotheses could then be assessed within the larger GIRD testing programme [12].

In order to achieve the goal of point (5), there is a need to firstly identify the type(s) of wear processes that may occur and then, from a more thorough understanding of the salient features of the various treatments, eliminate those that are unsuitable based on thickness, distortion, operating temperature, accuracy, flexibility, hardness considerations, etc. Further, rival processes can be eliminated on the basis of performance data, where applicable, and if necessary and where possible, by carrying out laboratory and then field trials. In this way, the accuracy and appropriateness of hypotheses (1–5) can be tested.

In practice, a large number of processes exist (see Table 13.3), and in many cases, it is clear that more than one acceptable treatment could be used. Cost and availability are then decisive. It may even be appropriate to use two (or more) techniques to impart optimum properties.

The Department of Metallurgy at the University of South Australia is presently following a similar procedure to identify possible viable alternatives to traditional wear-, and corrosion-resistant coatings and surface treatments used in the automobile industry [10]. These could include laser-hardening treatments, plasma nitriding, the range of modern nitrocarburising techniques, ion implantation, composite electroless coatings, and physical vapour deposition techniques. The objects of the present chapters are to review the advantages, limitations, and applications of the various processes on both commercial and technical grounds and to illustrate their use within the automobile industry with reference to published information collected during the recent study for Monroe, Australia [10].

13.2 SOME COATINGS AND SURFACE TREATMENTS: IDENTIFICATION, DESCRIPTION, AND APPLICABILITY

Methods of producing wear-resistant surfaces on metal components can be divided into two main categories: those infusion treatments where the hardened surface is produced by a heat treatment usually accompanied or preceded by diffusion of elements into the surface and those in which a coating or layer is formed or deposited on the surface (refer to Table 13.3). Table 13.4 provides a description or glossary of terms for the particular processes discussed within the chapter.

TABLE 13.3. Variety of Available Surface Treatments and Coatings.

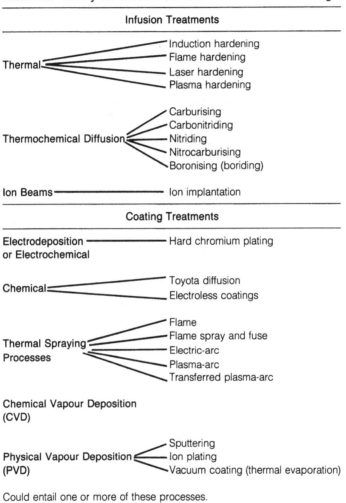

Infusion Treatments

Thermal
- Induction hardening
- Flame hardening
- Laser hardening
- Plasma hardening

Thermochemical Diffusion
- Carburising
- Carbonitriding
- Nitriding
- Nitrocarburising
- Boronising (boriding)

Ion Beams ——— Ion implantation

Coating Treatments

Electrodeposition ——— Hard chromium plating
or Electrochemical

Chemical
- Toyota diffusion
- Electroless coatings

Thermal Spraying
Processes
- Flame
- Flame spray and fuse
- Electric-arc
- Plasma-arc
- Transferred plasma-arc

Chemical Vapour Deposition
(CVD)

Physical Vapour Deposition
(PVD)
- Sputtering
- Ion plating
- Vacuum coating (thermal evaporation)

Could entail one or more of these processes.

TABLE 13.4. Description of Some Surface Coatings and Treatments.

Thermal Hardening Processes	Family of processes in which a suitable component surface is heated by a variety of methods above the upper critical transformation temperature and (normally) quenched to develop a hard martensitic structure.
Boronising	Boron, using either plasma, pack, paste, or salt processes is diffused into and combines with a metal substrate forming a hard metal boride layer. Process normally followed by hardening in vacuum or an inert atmosphere.
Nitrocarburising	Any of several processes conducted in a variety of processing media in which both nitrogen and carbon are diffused (absorbed) into surface layers of ferritic materials at temperatures less than the lower critical temperature ($<720\,°C$). Specific objective of forming nonmetallic compound layers and a nitrogen-rich subsurface diffusion zone on carbon or low alloy ferrous materials or components.
Toyota Diffusion (TD) Process	A proprietory process that produces pore-free compound layers of alloy carbides on carbon-containing ferrous materials. The carbide layer is produced by treatment at temperatures of between 800 and 1050°C in a fused borax bath containing ferro-alloys of the required carbide-forming element. Layer grows by reaction of the carbide former with carbon atoms diffusing from the substrate.
Electroless Nickel Coatings	Deposited by autocatalytic chemical reduction (electroless) of nickel ions by hypophosphite, aminoboranes or borohydride compounds. Can be modified for superior tribological properties by the addition of hard or high lubricity second phase particles and heat treatment.
Physical Vapour Deposition (PVD)	PVD encompasses a wide range of techniques in which the fundamental objective is to attract metal ions to the workpiece surface under the influence of an electrical bias and in the presence of a reactive gas under partial vacuum conditions. Process may be assisted by action of a slow discharge or plasma.
Chemical Vapour Deposition (CVD)	Deposition process in which a stream of gas containing volatile compounds of the element(s) to be deposited is introduced into a reaction chamber. Conditions within the chamber are controlled to enable an appropriate chemical reaction to take place, leading to formation of a coating on the substrate surface.

13.2.1 INFUSION TREATMENTS

13.2.1.1 Thermal Treatments

Of the wide range of thermal treatments used to impart surface hardening, *laser hardening* is considered to have the greatest growth potential, CO_2 gas lasers being the most suitable and most widely used for the majority of heat treatment requirements [5,9]. The process is normally self-quenching and the need for post-hardening operations is eliminated, thereby significantly reducing treatment time and post-operation machining. It is particularly suitable for large pieces and difficult-to-access surfaces. Other claimed advantages of the process [10,13] include

- superior process flexibility—ease of automation, low beam divergence giving localised and selective hardening control, and controlled penetration
- low heat output, high-speed efficient energy usage
- minimal distortion
- noncontact, chemically clean process

In other thermal treatments such as induction and flame hardening, the radiant energy, unlike that of the laser, is not coherent and hence cannot be transmitted through air over large distances; hence the heat distribution is much broader. These other processes may also require subsequent straightening or grinding operations to remove distortion and cracking caused by uneven heating, nonuniform cooling, and thermal shock during quenching [9]. Laser hardening also produces higher hardnesses (compared to, e.g., flame and induction hardening), suitable alloy chemistry, favourable residual stresses (fatigue resistance), fine microstructure, and smaller heat affected zones.

The process is now used in such diverse fields as automotive engineering, office machines, jet engines, personal products, printing, mining, drilling, etc.

Laser surface transformation hardening is extensively used in the automotive industry for hardening engine and drive-train components, such as cylinder liners, piston rings, crank shafts, and the like to reduce sliding wear and fatigue. Research [14] has shown that precision laser heat treatment of piston ring grooves of gray cast iron reduced the wear on piston rings as they moved in and out during the course of normal engine operation. Belfort [15] reported that laser hardening of the lobes of cast iron cams and fillets of crank shafts improved resistance to wear under high mechanical loads and velocity. Laser-hardened power steering gear housings, made of malleable ferritic ions, under high mechanical loads and velocity, were found to have a wear resistance nearly ten times greater than that of untreated castings [16]. In Reference [17], selective hardening (in the case of a steel shaft running in a bearing, using a CO_2 laser, and generating a spiral pattern of hard and soft zones) is said to result in one-third of the wear experienced by the uniformly induction-hardened shaft.

The advent of laser hardening has led to interest in applying electron beams for similar purposes. Unlike laser techniques, electron beam processes require vacuum conditions. On the other hand, such vacuum conditions provide an excellent environment to protect the heat-treated surface.

Clean metallic surfaces are, however, poor absorbers of radiation corresponding to the wavelength of CO_2 lasers, and it is generally necessary to cover the component with an appropriate coating, e.g., colloidal graphite.

13.2.1.2 Thermochemical Diffusion Treatments

Under heavy abrasive conditions, thicker coatings produced by the various *boronising* processes may be of use [9,18]. However, despite decreased coefficients of friction and improved corrosion resistance, the tooth-like nature of the coating/substrate interface is believed to impair both fatigue and impact strength of treated components.

The process has been further limited by the failure of commercial sources to promote the expensive boride carbon-based powders and pastes actively. Production costs are increased by the necessity for heat treatment to be carried out in an inert gas or vacuum in order to preserve the integrity of the boride layer. Subsequent heat treatment is necessary to produce monophased layers of Fe_2B in preference to the weak duplex layer of Fe_2B and FeB. The pack method does, however, avoid the use of expensive plant as any furnace can be used (and the component subsequently vacuum hardened if necessary in a contract shop). Further, plasma boronising offers the possibility of producing monophased boride layers during similar treatment cycles [19]. However, the process is still in the development stage, and a number of technical difficulties would have to be overcome to make it viable.

Plasma surface treatments such as plasma nitriding and plasma nitrocarburising [9,10,20,21] are worthy of particular consideration. Although they have been advocated as better and cheaper alternatives to ion implantation and laser alloying, higher capital costs compared with gaseous, salt-bath, and fluidised bed nitrocarburising have somewhat limited the applications of the techniques [10]. However, they do exhibit many benefits. For example, the process of plasma or ion nitriding offers the following advantages over its gaseous counterpart:

- No after-treatment quench is required.
- There is a savings in energy/gas consumption.
- There are greater process controls.

- Sputtering action prior to treatment removes oxide films, thereby further ensuring uniform layer thickness.
- The process initially depassivates workpiece surface—useful in high Cr tool steels that normally have patchy nitriding due to oxide films.
- The forced attraction of N ions to cathodic surface ensures more uniform case depths even for complicated shapes.
- There is no loss of surface finish.
- There is selective nitriding—simple mechanical shielding/masking/blanking.
- There is more uniform heating.
- Increased process rates lower nitriding temperatures, thereby reducing distortion and avoiding overtempering. High core strengths are obtained more easily: advantages for treatment of tools, dies, high-precision parts.
- There is more accurate control of nature and thickness of surface compound or white layer formation.
- At a given hardness level, the process produces a significantly more ductile diffusion zone.

Owing to the lower nitriding temperatures, the process can be used for a wider range of steels in line with their tempering characteristics. Further, as the processing temperatures are relatively low with no resultant transformations, distortion is low. Consequently, treated parts may often be used without further finishing operations.

Since finishing operations such as grinding may destroy favourable compressive stresses developed by case hardening, their elimination should improve the consistency of fatigue performance.

It is usual industrial practice, however, to nitride after an appropriate hardening and tempering operation. It is therefore essential that nitriding temperatures be below those of the prior tempering process.

For metal stamping dies, increased life/performance has been reported due to reduced scoring, galling, and pick-up whilst improving compressive strength and fatigue resistance. Friction coefficients are reduced sufficiently to aid material flow in the die, thereby saving on lubricant consumption [22]. Punches, plastic moulding dies, and car parts are regularly treated. Care is needed, however, in operation as plasma parameters have a marked effect on results, e.g., surface C content, case depth. Well-defined procedures for fixturing and masking are required.

Care also has to be exercised with regard to spacing on loading and the choice of process pressure to avoid "hollow cathode," which would result in nonuniform nitriding and localised overheating [23]. For best results, steels containing the nitride-forming elements such as Cr and V are necessary.

The range of patented gaseous, salt, or fluidised bed nitrocarburising techniques, owing to the favourable mechanical properties and wear characteristics they impart to metallic components, are being widely used in industry [24–27]. For example, salt-bath nitrocarburising techniques (e.g., Tufftride TF1) have been claimed to offer environmental advantages and generally low capital and operating costs. The process can also be automated in line with modern manufacturing trends. Further, salt quenching in the chosen AB1 salt solution appears to eliminate the need for detoxification of wash waters and concurrently produces an attractive black finish with enhanced corrosion resistance [26]. Nitroblack and the Nitrotec families of coatings, which include a post-nitriding oxidation step, offer similar improvements [22,24,28].

Stainless steels and high alloy steels can be difficult to nitrocarburise using gaseous techniques, owing to the formation of a thin surface layer of chromium oxide or other alloy oxides hindering the diffusion of nitrogen and reactants. It is possible to reduce this hindering effect by removing the oxide layer, e.g., by pickling or by shot blasting. These methods are not, however, always practical. A more convenient way to nitrocarburise such steels is to preheat the steel in an oxidising atmosphere, e.g., air. A thin porous iron-chromium oxide layer will then be formed on the surface. During the nitrocarburising process, the iron oxide will be reduced back to iron. The resulting oxide layer structure contains areas of iron that reach the surface, allowing the nitrogen and carbon to penetrate into the alloy.

However, most steels, including high-temperature resistant steels, stainless steels, cast irons, and high speed steels, can be successfully salt-bath nitrocarburised, although low alloy and plain carbon steels show the best results with regard to the thickness of the compound layer obtained. One major disadvantage that has been documented, however, is that in low cyanide salt-bath methods, the surfaces of the parts become severely oxidised and unsightly, especially with alloy steels.

13.2.2 COATING TREATMENTS

Coatings produced by the *TD process* (see Table 13.4) are reported to exhibit many beneficial properties, including high hardness and a significant corrosion resistance, at least in cases where chromium carbide is formed as a surface layer [29]. The simplicity of its operation makes it economically favourable to many more hi-tech processes and has, in fact, been advocated as a possible alternative to PVD in tooling applications where dimensional accuracy is *not* paramount. Coating adhesion is reported to be superior to that of hard chromium plate. The main limitation of the process, however, is the allowance that must be made for distortion during heat treatment and the narrow range of appropriate substrate compositions. The hardening/tempering-induced volume changes, it has been suggested, can be avoided by preheating, tempering, and then final machining, presuming the size tolerance is tight enough to justify this approach.

Electroless nickel coatings are finding increasing favour in engineering circles as both a wear- and corrosion-resistant material, although Ni-P is the only coating that has been developed extensively for commercial use [10,30]. The advantages of electroless nickel coatings include good coating/substrate adhesion, excellent uniformity and good throwing power, low labour costs, and the ability to be precipitation-hardened through low-temperature treatments. Such coatings have an acceptable resistance to corrosion and wear, although the optimised selection of an appropriate electroless coating requires a thorough knowledge of the relationships existing between process parameters; chemical, physical, and mechanical properties of the coatings; and their field behaviour. The wear resistance can, however, be improved by optimising heat treatment procedures [10,31–33].

The requirement for heat treatment to impart good wear resistance is nevertheless incompatible with good corrosion resistance as nickel phosphide particles precipitated during heat treatment deplete the matrix in phosphorous. Despite their appreciable wear-resistant properties, they cannot be used in applications where, in the absence of lubricant, seizure or galling can take place. In this respect modified electroless coatings such as electroless nickel/PTFE (e.g., "Niflor") offer significant promise, exhibiting lower coefficients of friction, increased lubricity, and favourable prices comparable to those of traditional electroless nickel coatings [34]. Low second phase particle concentrations are to be avoided if reduced fracture toughness and increased wear rates are to be avoided. Electroless nickel coatings containing hard second phase particles (of controlled size), such as SiC or Al_2O_3, also show some promise but are not yet commercially available [35].

Thin nitride or carbide coatings formed by physical vapour deposition (PVD) techniques, for example, possess many beneficial properties, including good adhesion, uniform thickness, low coefficients of friction, high hardness, and an aesthetically pleasing appearance requiring little or no post-coating finishing [10,36]. However, careful control of process conditions is essential if adherent protective coatings are to be consistently obtained, as discussed elsewhere [37].

Further, production costs reflect the need to have a large and predictable throughput and are most economic where jigging and process scheduling are constant. PVD processes are, however, economic for drills, gear cutting tools, forming tools, and dies. The cost of producing PVD-coated components presently precludes their use for mass-produced automotive parts although they are used extensively on racing cars.

Thin nitride or carbide coatings can also be produced by chemical vapour deposition (CVD). Although the process can be carried out at atmospheric pressure, operation at reduced pressures is recommended due to the benefits of improved coating uniformity and lower levels of contamination. The high hardness, low coefficient of friction coatings exhibits high wear resistance and extended component lives. However, treatment times can be long, process temperatures are high, and prolonged time at temperature can have a detrimental effect on mechanical properties due to grain growth and destabilisation of austenite [38]. In some cases, an expensive separate heat treatment carried out in vacuum in order to prevent coating oxidation is necessary. Further, as in the case of the TD process, allowance [39] must be made for distortion/dimensional changes during the CVD process and the subsequent hardening and tempering. CVD coatings also exhibit matte surfaces somewhat rougher than that of the substrate, whereas PVD and electroless coatings, for example, faithfully reflect the underlying surface [10,40]. As the CVD reaction takes place within a gaseous cloud, good coating uniformity can be obtained even with closely packed workpieces. The modified line-of-sight PVD process, on the other hand, generally has a more restricted throwing power, and workpieces must be loaded less densely in order to avoid "shadows" unless workpiece rotation is possible [40].

13.3 CONCLUDING REMARKS

It is evident from the preceding discussion concerning the applications, limitations, and advantages of the various processes that many cleaner, economic, and technically superior alternatives to "traditional" surface coatings and treatments do exist. Techniques such as electron beam or laser hardening, plus ion nitriding, would appear to be highly suited to the production of automotive parts with improved tribological properties. These techniques offer the advantages of greater process controls, more uniform surface properties and coating/layer thicknesses, ease of automation, elimination of post-operation machining (due to reduced distortion), and lower environmental problems. Other processes such as PVD may offer similar promise although, as yet, their unit cost precludes their use in the commercial car industry. Composite electroless coatings are also worthy of consideration. However, bearing in mind the sensitivities and complications associated with wear processes, it will be necessary to carry out laboratory and field trials to establish both optimum deposition/process conditions for all the various treatments in order to obtain acceptable wear performance. For example, it would be necessary for optimum adherence, porosity levels, and coating/layer thickness. Heat treatment procedures and component precleaning and preparation would also need to be carefully controlled.

13.4 ACKNOWLEDGEMENTS

The Department of Metallurgy at University of South Australia gratefully acknowledges the financial support of Monroe Australia Pty. Ltd, CSIRO, and University of South Australia.

13.5 REFERENCES

1 Peterson, M. B. "Introduction to Wear Control," *Wear Control Handbook,* ASME, NY, 1980.

2 Lansdown, A. R. and Price, A. L. *Materials to Resist Wear—A Guide to Their Selection and Use.* Pergamon Press, New York, 1986.

3 Czichos, H. "Failure Criteria in Thin Film Lubrication, the Concept of a Failure Surface," *Tribology,* 14–20, 1974.

4 Rigney, D. A. *Scripta Metallurgica et Materialia,* 24(15):799–803, 1990.

5 Ko, P. L. *Tribology International,* 20(2):66–78, 1987.

6 Ashby, M. F. and Lim, S. C. *Scripta Metallurgica et Materialia,* 24(5):805–810, 1990.

7 Almond, E. A. and Gee, M. G. *Wear,* 120:101–116, 1987.

8 Zum Gahr, K. H. Chapter 6, in *Microstructure and Wear of Materials.* Elsevier Publishers, 1987.

9 "Wear Resistant Surfaces in Engineering—A Guide to Their Production, Properties and Selection," Department of Trade and Industry Publication, HMSO, ISBN 0 11 5138269, 1986.

10 Cooper, A. R. and Strafford, K. N. "Coatings and Surface Treatments to Resist Wear and Corrosion of Automobile Parts," Stage 1 report on Techsearch/University of South Australia Project 1935, August 1990.

11 Cooper, A. R., Strafford, K. N. and Sare, I. "Selection Criteria and Design Guidelines for Surface Treatments and Coatings for Wear Resistance," University of South Australia/CSIRO Collaborative Research Grant.

12 "Evaluation of Advanced Surface Coatings," DITA/GIRD Project, Department of Metallurgy, University of South Australia.

13 Eckersley, J. S. In *Advances in Surface Treatments—Technology, Application, Effects. Volume 1,* A. Niku-Lari, ed., Pergamon Press, pp. 211–231.

14 Seaman, F. D. and Gnanamuthu, D. S. *Metal Progress* (August):210–215, 1975.

15 Belforte, D. A. In *Colloqium on Lasers and Electro-Optical Equipment.* Tokyo, Japan, 1978.

16 Miller, J. E. and Wineman, J. A. *Metal Progress* (May):38–43, 1977.

17 *Lasers—Operation, Equipment, Application and Design* (Coherent Inc.). McGraw-Hill, New York, pp. 124–135, 1980.

18 Dearnaley, P. A. and Bell, T. *Surface Engineering,* 1(3):203–217, 1985.

19 Bloyce, A., Dearnley, P. A. and Bell, T. In *Proceedings of 1st International Conference on Surface Engineering,* Paper 37, Volume 1, Brighton, England, June 1985, Bucklow, I. A., ed., Crampton and Sons Ltd. Publishers. Sponsored by Surface Engineering Society and the Welding Institute.

20 Edenhofer, B. *Heat Treatment of Metals,* 1:23–28, 1974 and 2:59–67, 1974.

21 Childs, H. C. *Heat Treatment of Metals,* 3:73–75, 1986.

22 Hick, A. J. and Close, D. "Report on 11th ASM Heat Treating Conference and Exposition/6th International Congress on Heat Treatment of Materials," *Heat Treatment of Metals,* 1:13, 1989.

23 Dixon, G. J., Plumb, S. A. and Child, H. C. "Processing Aspects of Plasma Nitriding," *Heat Treatment 81,* Book No. 283, The Metals Society, pp. 137–146, 1983.

24 Dawes, C. and Tranter, D. E. *Heat Treatment of Metals,* 3:70–76, 1985.

25 Etchells, I. V. *Heat Treatment of Metals,* 4(8):85–88, 1981.

26 Lasday, S. B. *Industrial Heating* (July):22–23, 1988.

27 Eysell, F. W. *Machine Tool Engineering and Production,* 98(5):109–113, 1985.

28 Dawes, C. and Tranter, D. F. *Metal Progress,* 124:(8)17–22, 1983.

29 Jackson, P. T. and Child, H. C. *Heat Treatment of Metals,* 4:92–94, 1985.

30 Strafford, K. N., Datta, P. K. and Gray, J. S. Chapter 3.1.1 in *Surface Engineering: Fundamentals, Processes and Applications.* Ellis Horwood, Chichester, UK, 1990.

31 Gawne, D. T. and Ma, U. *Surface Engineering,* 4(3):239, 1988.

32 Gawne, D. T. and Ma, U. *Wear,* 120:125–149, 1987.

33 Datta, P. K., Strafford, K. N. and Allaway, S. Chapter 3.4.3 in *Surface Engineering: Fundamentals, Processes and Applications.* Ellis Horwood Press, Chichester, UK, 1990.

34 Ebdon, P. R. *Int. Journal Materials and Product Technology,* 1(2):290–300, 1986.

35 Boose, C. A. *Proc. Final Contractors Meeting on European Research on Materials Substitution,* Brussels, pp. 253–258, Elsevier Appl. Science Publishers, London, 1986.

36 Dearnaley, P. A. *Heat Treatment of Metals,* 4:83–91, 1987.

37 Ahmed, N. A. G. *Ion Plating Technology—Developments and Applications.* John Wiley, New York, 1987.

38 Perry, A. J. and Horvath, E. *Metals and Materials* (October):37–40, 1978.

39 Vagle, M. *Journal of Heat Treating,* 2(2):121–129, 1981.

40 "Surface Treatment of Tools and Dies—The Options," *Heat Treatment of Metals,* 3:77–83, 1983.

CHAPTER 14

Characterisation and Quality Assurance of Surfaces and Surface Coatings

H. WEISS[1]

14.1 INTRODUCTION

MODERN designs of components for the power-generating, aircraft, or space industry place such high demands on the materials specified that they can very often only be met by tailoring composite materials for these specific applications. In particular, the requirements for bulk properties, on one hand, and surface properties, on the other hand, differ so much that the surfaces have to be specially treated and modified to meet the particular demands. This treatment is termed *surface engineering,* and in order to meet the specifications, the surfaces have to be characterised and the methods of quality assurance have to be applied similar to the procedures used with the bulk material.

The surface characterisation, with respect to composition, microstructure, and properties, is the decisive step in research and development of surface treatments because the outcome determines whether or not specific design requirements can be achieved (Figure 14.1). It can be done in a destructive or nondestructive way, and it is also an important feature of quality assurance, once the particular surface engineering treatment has gone into production. But surface characterisation is, by no means, the only way of assuring quality. As in the production of bulk materials such as steel, process parameter control, which is the control of routines found to yield the desired product, is the foremost procedure; the monitoring of components' behaviour in service also serves to assure the functioning of the surface.

[1]Laboratory of Surface Technology, Siegen University, P.O. Box 101 240, D-5900 Siegen, Germany.

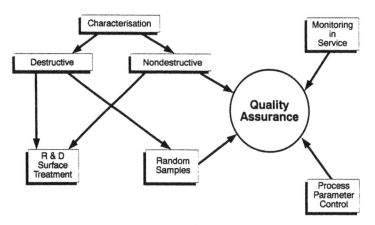

Figure 14.1 Characterisation as part of the quality assurance procedures.

All these different procedures serve to develop and produce surfaces tailored for specified loadings and functions, and they will be discussed in detail in this chapter.

14.2 CHARACTERISATION

Generally, surface engineering can be carried out in three different ways (Figure 14.2):

- preparing the surface of bulk materials without changing its

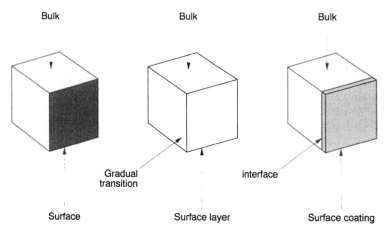

Figure 14.2 The different types of surface engineering.

composition or microstructure, e.g., by polishing to increase the corrosion resistance or the bearing capacity of slide bearings
- modification of the surface, e.g., by cold working, diffusion processes, etc., without creating a well-defined interface
- coating the surface with a different material, which incorporates an interface

Obviously, in the first case, it is sufficient to characterise function and structure of a few atomic layers; in the second case, the entire modified zone has to be taken into account, and in the third case, the problems of the interface also have to be dealt with.

As far as characterisation and quality assurance are concerned, the third case incorporates the two other cases as well, although with modified surface layers compositional, structural, and property gradients come into account. However, in these considerations, the case of surface coatings will generally be discussed.

The system involving bulk material, interface, surface coating or modified surface layer, and surface is very complex and, together, yields the system's properties, which are requested by the designer, with every part playing an important role. The complexity of the system is shown in Figure 14.3.

It is possible to distinguish between loading and functions of a surface. While loading requires resistive properties of the surface to withstand these conditions, the latter involves "active" surface properties, e.g., the property to reflect certain wavelengths of the electromagnetic spectrum. As far as properties are concerned, one can differentiate between properties that involve only the top atomic layers, such as surface energy, or properties such as hardness, where a surface layer of at least several micrometers is involved.

In the case of a modified surface layer, composition and properties exhibit gradients over the thickness of the layer, and the interface is absent.

Surface characterisation means the quantitative assessment of the relevant properties by means of physical, chemical, and technological effects. Here it is practical to distinguish between the characterisation with respect to structure and composition and the characterisation with respect to the other properties.

14.2.1 STRUCTURE AND COMPOSITION

One can use a great number of interactions to charactierise a surface or a metallographic section (Figure 14.4). Feeler gauges can be used to study the surface contours, or gases and liquids to study porosity and surface energy. Notably, the interactions of a surface with photons and electrons not

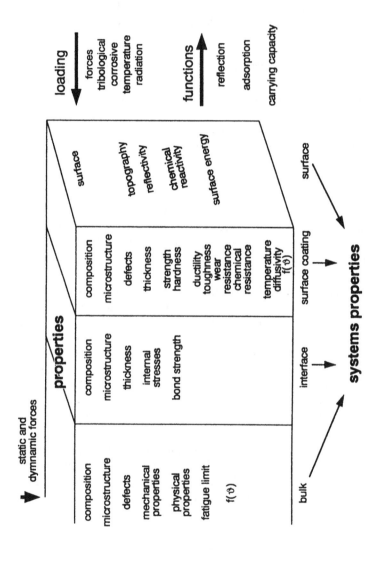

Figure 14.3 Properties of a component as a result of interactions.

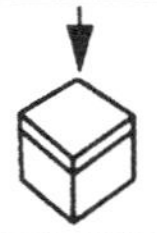

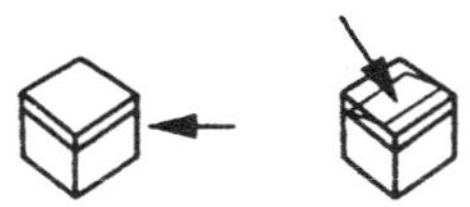

	topography	reflectivity	surface energy	micro-structure	internal stresses	composition	defects
feeler gauge							
gases	roughness			tunneling microscopy			
liquids	BET		dihedral angle				
ions						SIMS GDOS, ISS	
electrons	imaging SEM, TEM					EDX, WDX AES	
photons	imaging holography	gloss colour IR-spectr.		X-ray diffraction	X-ray diffraction	XPS, UPS X-ray fluor.	
US waves							reflection US microsc.

electrons				imaging SEM, TEM		EDX, WDX	
photons				imaging X-ray diff.		X-ray fluor.	

Figure 14.4 Characterisation of treated surfaces with respect to structure and composition.

only lead to images via light and electron microscopy, but also yield information on chemical composition, molecular structure, and binding status; by way of X-ray diffraction, knowledge of lattice parameters can be obtained.

Figures 14.5 and 14.6 show examples of a coating with detachment at the interface (light microscopy) and a surface with cracks and erosion traces (SEM).

Examples of modern imaging are the holographic characterisation of the surface profile of an object [1] (Figure 14.7), which corresponds very well with the feeler gauge profiling and the highly developed methods of surface analysis. In Figure 14.8, the information depths of different analytical methods are compared [2], namely, interaction with electrons yielding X-rays (EDX), interaction with electrons or photons yielding auger and photo electrons (AES, XPS), interaction with ions yielding secondary ions (SIMS), and interaction with ions leading to scattering those ions (ISS). Obviously, only the latter three methods are capable of characterising the immediate surface, EDX actually measuring the builk of the material. By removing the surface layer by layer via sputtering or ball-cratering, the

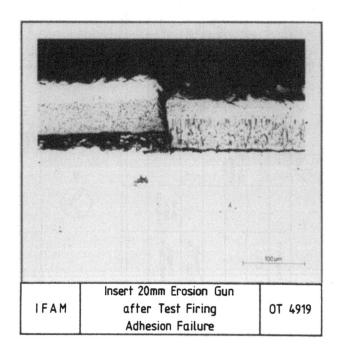

| IFAM | Insert 20mm Erosion Gun after Test Firing Adhesion Failure | OT 4919 |

Figure 14.5 Adhesive failure of a Cr-coating tested by a hot gas pulse (light micrograph of a section).

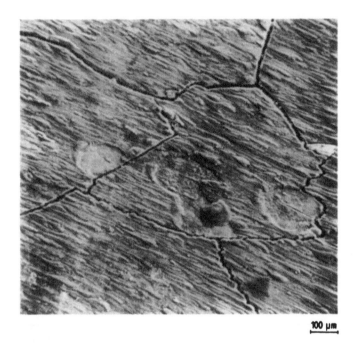

Figure 14.6 Tested surface showing cracking and particle erosion traces (SEM).

surface analysis methods can be used to obtain depth profiles of surface layers; Figure 14.9 presents an example of a 30% Cr and 70% Fe–containing coating, showing the oxide on top of the coating, as well as the adsorbed carbon.

14.2.2 SPECIFIC PROPERTIES

The properties of the surface layer or coating are also obtained by interactions. The interaction with a diamond pyramid or indentor yields results on the hardness of the area measured, and interactions with gases, liquids, particles, or other sufaces yield information on hot corrosion, electrolytic corrosion, erosion, or wear resistance (Figure 14.10). Measuring hardness, microhardness, or penetration by nano-indentor is fairly straightforward, although the stress status is complex. Measurements of the other properties are extremely involved due to the almost infinite number of parameters. This has resulted in a great variety of tests, which cannot be dealt with here. Determining wear resistance alone, which can be classified in abrasive wear, sliding wear, and rolling wear [3] (Figure 14.11) has led to the development of a great number of tests, which all serve

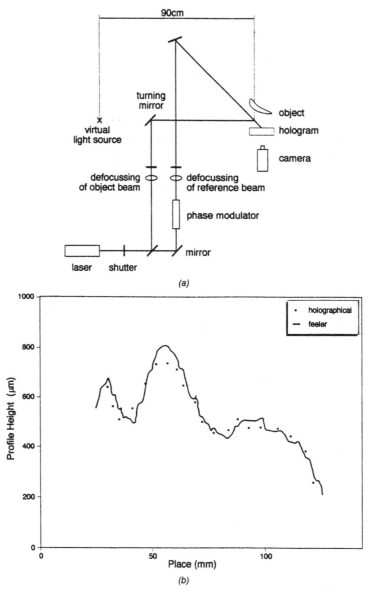

(a)

(b)

Figure 14.7 (a) Holography arangement to determine surface profiles; (b) Comparison of a profile obtained by holography and by feeler gauge.

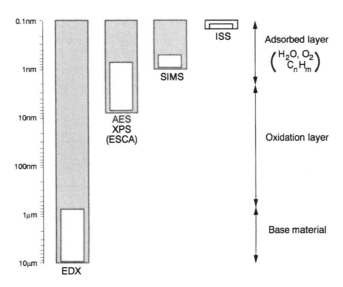

Figure 14.8 Information depth of different methods of analysis [2].

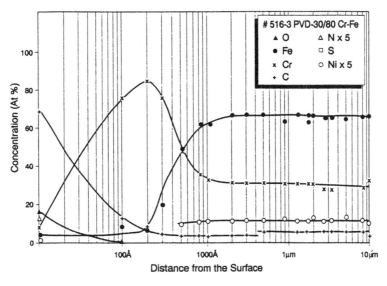

Figure 14.9 AES depth profile of a PVD coating (30 at.% Cr; 70 at.% Fe).

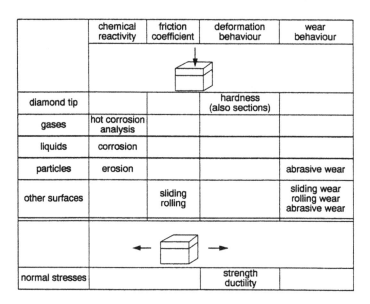

	chemical reactivity	friction coefficient	deformation behaviour	wear behaviour
diamond tip			hardness (also sections)	
gases	hot corrosion analysis			
liquids	corrosion			
particles	erosion			abrasive wear
other surfaces		sliding rolling		sliding wear rolling wear abrasive wear
normal stresses			strength ductility	

Figure 14.10 Interactions to characterise surfaces with respect to specific properties.

Wear Testing

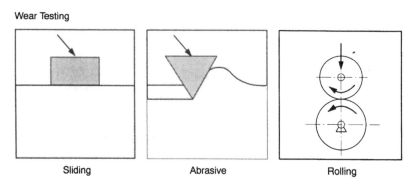

Sliding   Abrasive   Rolling

Figure 14.11 The different types of wear as base for wear testing arrangements [3].

226

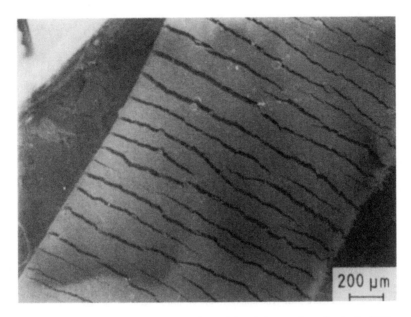

Figure 14.12 Crack pattern of a Cr-coating evolving during tensile testing in the SEM.

a particular purpose, reflecting the fact that wear resistance is a system property.

By applying normal stresses to specimens, the ductility of the surface layer can be assessed by the crack pattern that evolves. As an example, Figure 14.12 shows a chromium-coated tensile specimen tested in a scanning electron microscope in tension.

14.2.3 BOND STRENGTH

The interfacial properties influence the behaviour of coatings under loading, and it is of particular interest to study the strength of the bond between coating material and substrate. Due to the thinness of coatings, which generally varies between 500 μm for sprayed coatings and less than 10 μm for hard coatings, this task is very difficult. The tests available either use normal stresses, shear stresses, or a fracture mechanical approach (Figure 14.13). The tension tests suffer because a counter bar has to be bonded to the coating either by adhesives or by brazing. By this procedure, the coating and the interface itself might be affected, and very often, the adhesive bond is too weak to test a well-adhering coating. The shear tests suffer from the very exacting and costly sample preparation and cannot be applied to coatings less than 100 μm thick. In the case of fracture

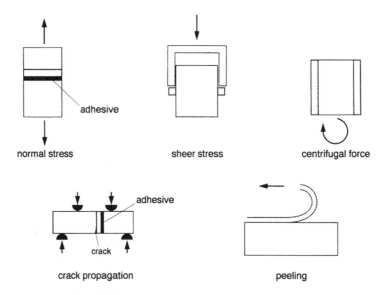

Figure 14.13 The different ways of measuring bond strength.

mechanical testing, the introduction of cracks is extremely involved. The centrifugal force method is difficult to control, and peeling is restricted to polymer coatings. Summarising, the quantitative characterisation of bond strength has not yet been solved successfully.

14.2.4 COATING THICKNESS

As it is apparent from Figure 14.14, the coating thickness can be measured in many different ways, depending on the properties of coating and substrate. Without going into details, it can be stated that there is a thickness measuring method for any coating on any substrate, yet not necessarily nondestructive.

14.2.5 NONSPECIFIC TESTING

So far, characterisation methods have been discussed, which measure specific properties in a quantitative way. However, there are a great number of nonspecific tests available that have evolved during the practical work on coating/substrate systems. Generally, they apply a very complex loading profile and yield results on bond strength, cohesive strength within a coating, bonding defects, and ductility of a coating, and they will often work very well assessing the general quality of a coating and its bonding to the substrate. Some of them are listed in Figure 14.15. Notably, the

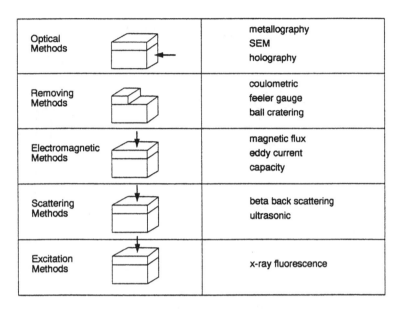

Optical Methods		metallography SEM holography
Removing Methods		coulometric feeler gauge ball cratering
Electromagnetic Methods		magnetic flux eddy current capacity
Scattering Methods		beta back scattering ultrasonic
Excitation Methods		x-ray fluorescence

Figure 14.14 Different methods of measuring the coating thickness.

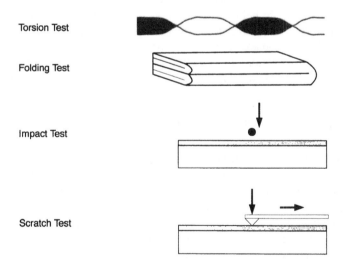

Torsion Test

Folding Test

Impact Test

Scratch Test

Figure 14.15 Tests with complex loading profiles (nonspecific tests).

229

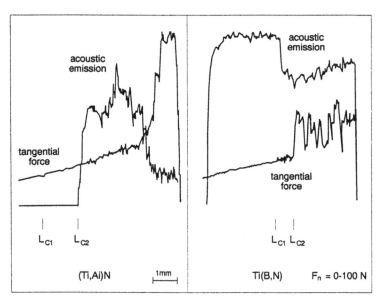

Figure 14.16 Comparison of scratch tests on (Ti,Al)N and Ti(B,N) coatings: L_{C1} = cohesive failure; L_{C2} = adhesive failure [4].

scratch test is very widely used to assess the quality of hard coatings. It involves a diamond tip being moved across the coating in test and increasingly loaded by a normal force. During the test, the tangential force and the acoustic emission are recorded and irregularities of the force or acoustic bursts are indications of cohesion or adhesion of the coatings being scratched (Figure 14.16). The interpretation of results requires great experience.

14.2.6 TESTING UNDER SERVICE CONDITIONS

Most surface-engineered components are subjected to very special and often complex loading profiles in service, which is particularly the case in corrosive aqueous and gaseous environments and for tribological applications. In order to obtain data on the actual performance in service, the components have to be tested under closely simulated service conditions. Burner rig testing of turbine parts and corrosion testing under atmospherical or seawater conditions are examples of this type of surface characterisation.

The spectrum of tribological testing is particularly extensive and comprises very cheap pin/disk arrangements, on the one hand, and extremely expensive field testing, on the other hand. Figure 14.17 illustrates this testing philosophy [5].

14.3 QUALITY ASSURANCE

Once a surface treatment has gone into production, quality assurance procedures have to be applied to maintain the standard achieved during development work and be considered adequate for the component. These procedures are of the same nature as the ones used during the production of bulk materials and can be categorized as follows:

- process parameter control
- random sampling
- nondestructive testing
- monitoring in service

14.3.1 PROCESS PARAMETER CONTROL

This most important of procedures deals with the control of all raw materials and processes used during the surface treatment. As an example, Figure 14.18 shows the parameters that influence the coating quality of air plasma-sprayed coatings. These parameters have to be strictly controlled and maintained within specific limits in order to obtain uniform quality over a product series and periods of time. It is well known that, if a quality problem arises, the first remedial step has to be the checking of any deviations in the process parameters.

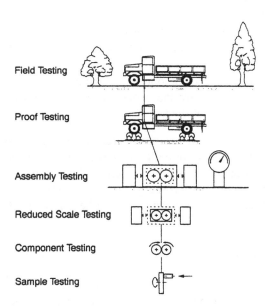

Figure 14.17 Spectrum of tribological testing [5].

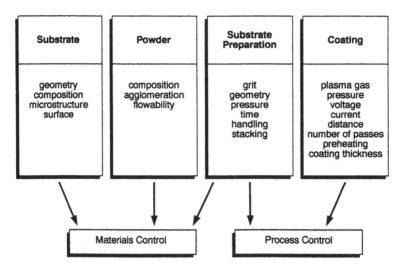

Figure 14.18 Important process parameters for air plasma spraying.

14.3.2 RANDOM SAMPLING

The testing methods described to characterise surfaces and coatings are all being used for quality assurance as well, depending on type and purpose of the particular component and the particular surface treatment. One can distinguish between two different cases:

(1) In the case of small components, one or several samples are taken from every batch produced and tested with respect to the relevant properties.

(2) In the case of larger components, separate small samples are often prepared under exactly the same process conditions and then subjected to the tests selected.

In both cases, destructive and nondestructive testing methods can be distinguished by whether or not the particular component is destroyed either by the test itself or by cutting out samples. Thus, a number of tests such as scanning electron microscopy or micorhardness measurements, although nondestructive by nature, have to be classified "destructive" for larger components.

14.3.3 SPECIAL NONDESTRUCTIVE TESTING METHODS

In Figure 14.19, some of the most common nondestructive testing methods are listed. Among them are those of particular interset that allow testing of every single piece (100% testing). The most common one used is the simple visual inspection of the treated pieces and categorizing them according to certain standards. Coating defects, e.g., blisters, uncoated areas, and changes in colour can be detected quite easily this way. As an example, the colour of TiN coatings depends very much on the chemistry, and a bright golden surface indicates the stoichiometric composition of TiN. Most coating thickness measuring methods can be used for 100% examination; the breakdown voltage is a good indicator of the density/porosity of an insulating coating.

Ultrasonic (Figure 14.20) and holographic methods [1] are very interesting, but as far as their application is concerned, they are still in their infancy.

14.3.4 MONITORING IN SERVICE

Quality assurance not only means assuring the production quality before delivery, but also implies that surfaces and suface coatings may not fail when in use, which could often result in expensive or catastrophic side effects. One way of monitoring the quality of components like bearings is the spectrographic analysis of the metal contents of the engine oil. As an example, Figure 14.21 depicts the large increase in the Fe content in the oil of a jet engine, indicating a faulty bearing [6].

Other quality indicators are temperature rises or increases of the sound

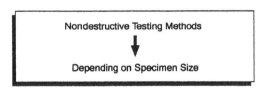

Figure 14.19 Nondestructive testing methods.

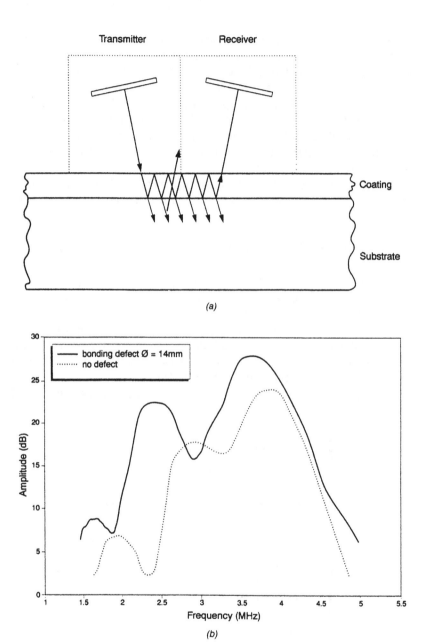

(a)

(b)

Figure 14.20 Ultrasonic testing of a coating for bonding defects [l]: (a) testing arrangement; (b) results.

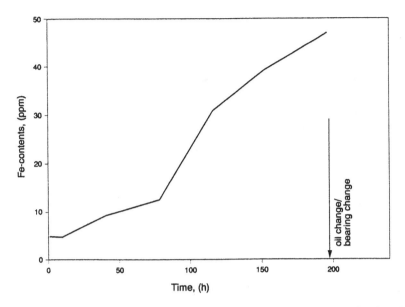

Figure 14.21 Surface quality monitoring by analysis of the oil of a jet engine [6] (after Orcutt).

emission of bearings (Figure 14.22). By the frequency range emitted, the different bearings become distinguishable, and an amplitude limit can be defined beyond which the surface of a bearing is faulty and needs to be replaced [6].

14.4 EXAMPLES OF CHARACTERISATION AND QUALITY ASSURANCE PROCEDURES

14.4.1 HARD COATINGS FOR TOOL TIPS

Quality control of hard coatings on tools or tool tips presents a difficult problem because the coatings are thin, less than 10 μm, and coating and adhesive properties have to be excellent if the coating is not to fail during use. In Figure 14.23, the relevant properties are listed together with the testing methods that can be applied [4]. Interestingly, the scratch test, with its undefined loading profile, has to be used in order to measure the adhesive and cohesive properties. Measuring the coating thickness also presents a problem and can only be done destructively. In Figure 14.24, the quality assurance procedures are listed [4]. Only visual inspection and feeler gauge can be used for a 100% inspection; the other testing methods can only be applied to random samples.

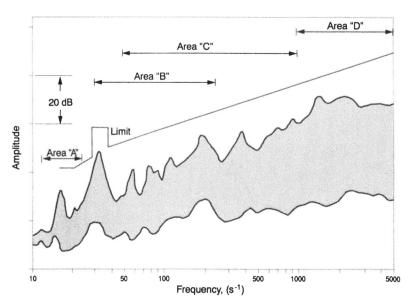

Figure 14.22 Sound emission of submarine pump bearings [6] (after Bowen and Graham).

Testing method				Property
Metallography (MET)		Scanning electron microscopy (SEM)		Structure
MET	SEM		feeler gauge (FG)	Topography
MET	SEM	FG	ball cratering	Thickness
Scratch test				Hardness
Scratch test				Adhesion Cohesion
Pin/disc		Ball/disc		Friction wear

Figure 14.23 Possible testing sequence for hard coatings [4].

236

Testing method	All coated parts	Random sampling	Random samples after process change
Visual inspection	+	+	+
metallogr. section	-	o	+
SEM	-	o	+
feeler gauge	o	+	+
ball cratering	-	+	+
scratch test	-	+	+
pin/disk or ball/disk	-	o	+

+ suited	o partly suited	- not suited

Figure 14.24 Quality assurance procedure for hard coatings [4].

The bond strength is very sensitive to the composition of the interface, and adhesive failures can often be traced to contamination of the substrate surface before coating. Depth profiling by sputtering and auger electron spectroscopy are very powerful ways to determine such contaminations. This is shown in Figure 14.25, where a cohesion defect could be traced to an enrichment of carbon at the interface.

14.4.2 TURBINE BLADES

Turbine blades are often coated with a complex coating system, comprising a MCrAlY bond coat on the substrate, a thermal barrier coating of a stabilized ZrO_2, followed by another MCrAlY hot corrosion-resistant coating. Additionally, varying coating thicknesses are often demanded.

Apart from other more conventional characterisation methods, two modern interesting testing procedures can be applied. The coating thickness can be determined by holographically imaging the profile before and after coating [1] (Figure 14.26), and the temperature diffusivity of the coating system can be measured by an energy flash method, where the time elapsing between a heat pulse and its detection at the back of the sample is measured (Figure 14.27).

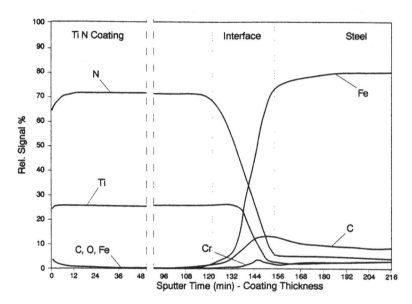

Figure 14.25 Depth profile of a TiN coated HSS drill showing enrichment of carbon at the interface.

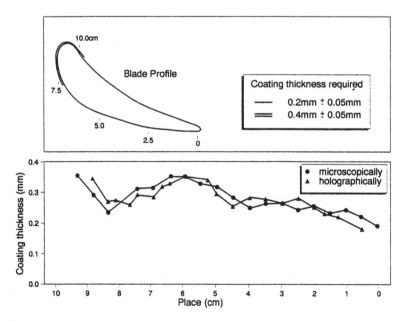

Figure 14.26 Determination of the coating thickness by holography and metallography. In the area between 7 and 10 cm, the coating is too thin compared to the specification [1].

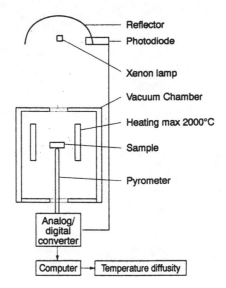

Figure 14.27 Arrangement to measure temperature diffusivity by the heat pulse method.

14.4.3 SMOOTH BORE GUN BARRELS

Gun-barrel erosion is a type of damage that determines the service life of a high-performance gun. Figure 14.28 shows a typical example of this type of surface deterioration. For coating development, as well as for quality assurance, small test pieces are coated and tested in an arrangement where the conditions of a round can be simulated as closely as possible. The testing arrangement consists of a pressure vessel, which is supported by a large concrete construction to take up the stresses. Test pieces are mounted in a way to form an exit vent for the propellant gases (Figure 14.29). The test pieces are characterised after testing with respect to erosive wear and reaction layers formed. Depth profiling by AES yields information on the erosive mechanisms, e.g., the formation of a carburized zone with low melting point (Figure 14.30).

14.5 CONCLUSION

It can be concluded that surface and surface coating characterisation, as well as quality assurance, are very important parts of surface engineering. A great variety of powerful testing methods exists both to characterise surfaces and coatings and to ensure that the quality is adequate. Nondestructive coating methods, however, that can be used for 100% testing are still in the development stage, and further work has to be done in this area.

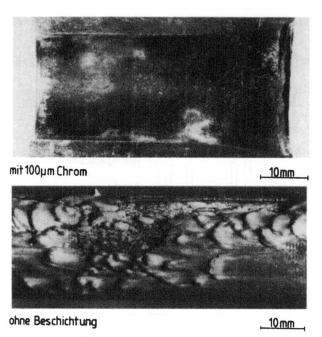

mit 100 μm Chrom   10 mm

ohne Beschichtung   10 mm

Figure 14.28 Sectioned gun barrel showing erosion damage (a) and positive effect of 100 μm Cr-coating (b).

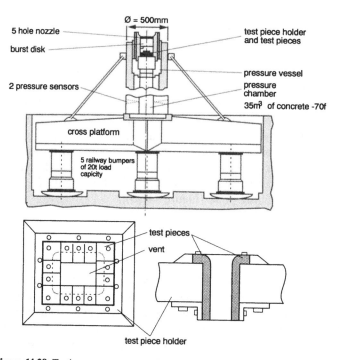

Figure 14.29 Testing arrangement to simulate erosive conditions in a gun barrel.

240

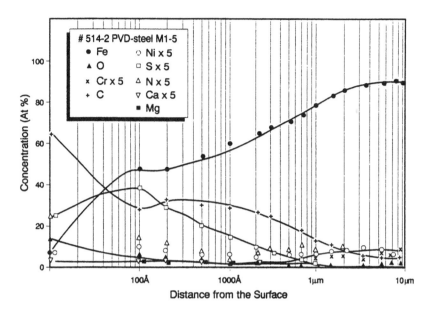

Figure 14.30 AES depth profile of a PVD-steel coating tested with a propellant. Note the carburized reaction zone formed.

14.6 REFERENCES

1 Crostack, H.-A., Krüger, A. and Pohl, K.-J. "Zerstörungsfreie Qualitätssicherung von Beschichtungen," in *Beschichtungen für Hochleistungsbauteile, Symposium Hagen, 6. und 7.*, VDI-Berichte 624, VDI-Verlag, Düsseldorf, 1986.

2 Hantsche, H. "Zum Einsatz von Oberflächenanalyseverfahren zur Unterstützung von tribologisch beanspruchten Werkstoffoberflächen," in *Reigung und Verschleiß bei metallischen und nichtmetallischen Werkstoffen*, K.-H. Zum Gahr, ed., DGM-Informationsgesellschaft, Oberursel, 1990.

3 Zum Gahr, K.-H. "Metallische und Keramische Werkstoffe," in *Reibung und Verschleiß bei metallischen und nichtmetallischen Werkstoffen*, K.-H. Zum Gahr, ed., Oberursel, 1986.

4 Matthes, B., Broszeit, K., Kloos, H. "Prüfung und Qualitätssicherung von PVD-Hartstoffbeschichtungen," in *Reibung und Verschleiß bei metallischen und nichtmetallischen Werkstoffen*, K.-H. Zum Gahr, ed., DGM-Informationsgesellschaft, Oberursel, 1990.

5 Czichos, H. "Stand und Perspektiven der Tribologie," in *Reibung und Verschleiß bei metallischen und nichtmetallischen Werkstoffen*, K.-H. Zum Gahr, ed., Oberursel, 1986.

6 Habig, K.-H. *Verschleiß und Härte von Werkstoffen*. Carl Hanser Verlag, München, Wien, 1980

CHAPTER 15

Surface Engineering and the Control of Corrosion and Wear

K. N. STRAFFORD[1]
P. K. DATTA[2]

15.1 INTRODUCTION

STRAFFORD, Datta, and Gray [1] have provided an outline of the subject of surface engineering, whose definition and scope, it is suggested, encompasses techniques and processes used to modify and enhance the performance of surfaces with respect to wear, fatigue, corrosion resistance, and biocompatability. In particular, they have identified three constituent interactive areas (Table 15.1), namely,

(1) Optimisation of surface properties
(2) Surface treatments and coatings technology
(3) Characterisation of coatings

In area 1, research concerns the optimisation of the performance of both surfaces and substrates, i.e., the coating system in terms of corrosion, wear, fatigue, adhesion, and other physical and mechanical properties. A whole variety of processes and treatments exist within area 2, embracing both what are basically old technologies (but ones that are often undergoing continuous development to meet competition and materials substitution), as well as the new advanced surface engineering technologies such as plasma-assisted carburising or nitriding or physical vapour deposition (ion plating).

Failures in mechanical systems occur by the processes of corrosion, wear, and fatigue, each with its own particular characteristics. Corrosion

[1]Department of Metallurgy, University of South Australia, SA 5095, Australia.
[2]Department of Mechanical Engineering and Manufacturing Systems, University of Northumbria, Newcastle-upon-Tyne NE1 8ST, UK.

243

TABLE 15.1. **Surface Engineering: Definition and Scope.**

- Encompasses techniques and processes used to induce, modify and enhance performance of surfaces with respect to:
  —Wear
  —Fatigue
  —Corrosion Resistance
  —Biocompatibility
- Three major categories of interrelated activities:
  —Optimisation of surface properties
  —Surface treatments and coatings technology
  —Coatings' characterisation

Optimisation of Surface Properties

Research concerns the optimisation of the performance of surfaces and substrates, i.e., the coatings system—corrosion, wear, fatigue, adhesion, and other physical and mechanical properties.

Surface Treatments and Coatings Technology

Painting, electroplating, electroless deposition, weld surfacing, plasma and hypervelocity spraying, thermochemical treatments—nitriding and carburising, laser surfacing, physical and chemical vapour deposition, and ion implantation.

Coatings Characterisation

Composition, morphology, structure, electrical and optical properties.

may be defined as deterioration by chemical or electro-chemical reaction with the surrounding environment. Wear involves the loss of material and damage to objects and machinery due to mechanical contact. Wear, although rarely catastrophic, is a particularly complex process, and five distinct forms are recognised [2]—adhesive, abrasive, chemical, fretting, and erosive—each varying in incidence in the engineering world. Wear reduces operating efficiency through increasing power losses, oil consumption, and the rate of component replacement.

The processes of corrosion, wear, and fatigue are all expensive, with direct and indirect cost elements. It is now well recognised that the successful exploitation of these various surface treatments and coatings technologies may allow the use of simpler, cheaper substrate or base materials, with concomitant substantial reduction in costs, minimisation of demands for strategic materials, and improved production and performance. In other, more exacting situations, the use of specially developed coatings may represent the only real possibility for exploitation of a process technology. This technology may also allow the use of rare or scarce, and therefore expensive, materials, which are known to possess novel properties.

The essential and central role of surface engineering and coatings technology in design and materials selection is identified and elaborated on in Table 15.2. Here, a crucial point is the need to consider surface engineering ab initio in the design process; it should not be seen as a last resort to meet particular engineering surface problems.

Table 15.1 identifies a number of surface treatments and coating technologies currently used to enhance surface properties. Some of those processes have been in use for many, many years, e.g., the coating of steel with zinc ("galvanising"); others are very new technology indeed—physical vapour deposition (PVD) has only been pursued over the last decade and is the subject of intensive research at the current time. The historical dimensions of surface engineering are emphasized further in Table 15.3, where two central features are noteworthy: firstly, the manner in which traditional technologies are continually being improved and upgraded, and secondly the advent of entirely new technologies. "Fitness for purpose," however, is the overriding consideration, within the constraints of the overall design audit.

The object of this chapter is to consider the processes of corrosion and wear and to indicate the use and role of surface coatings and treatments, i.e., surface engineering in the control of these modes of degradation. The chapter is illustrated with reference to recent and ongoing research.

15.2 SURFACE ENGINEERING AND CORROSION CONTROL

Some of the key aspects of the problem of corrosion and the rationale whereby surface engineering can be used with advantage to control corrosion are summarised in Table 15.4. It has been convenient to consider the use, potential, and role of surface engineering for corrosion control under

TABLE 15.2. Surface Engineering (SE) in Design and Materials Selection.

- *Optimisation* of benefits to be achieved from SE necessitate it to be considered as an integral part of the design audit.
- SE, in allowing the choice of desired surface properties independently of bulk component properties, provides considerable flexibility in terms of materials selection.
- SE has removed the constraint to manufacture components from very high cost bulk materials to take advantage of unique surface priorities, since virtually any material can now be deposited as a continuous thin film.
- While surface properties can be engineered, the surface must be adequately supported by the substrate: hence, the need to consider the total design audit/requirement of the coatings/surface treatment substrate system.
- SE must be considered ab initio at the design stage in order to be able to optimise performance: it should not be used to solve or alleviate service problems caused by poor materials or bad design.

TABLE 15.3. The Development of Surface Engineering:
A Historical Perspective.

- SE is basically an old practice or technology.
- The continually increasing demands made of materials with enhanced properties have added impetus to the development of sophisticated new techniques as well as equally significant enhancement of existing practices.
- Numerous examples may be cited:
  —the advent of plasma-assisted carburising and nitriding in the heat treatment of steels for enhanced cutting and wear performance
  —the continuous development of zinc coatings, for improved corrosion resistance, with enhanced fabricability of coated sheet steel
  —the replacement of electroplated nickel by electroless nickel coatings, and the prospects for corrosion resistance coupled with improved tribological performance
  —the rapid development of the newer vapour deposition processes—both CVD and especially PVD—to allow the true design of, and optimisation in, coatings systems, especially in hybrid or multilayer form.
- SE, fully developed, will pave the way for real technological advances where the prospect of total "fitness-for-purpose" in engineering situations may be the more nearly achieved.

TABLE 15.4. Surface Engineering and Corrosion.

Corrosion	• one of the most important, wide-ranging, performance-limiting technical problems to be faced by industry
	• hugely expensive, e.g., cost to UK economy, £6,500M per year
	• common cause of premature component failures
	• introduces price penalties by necessitating the use of expensive, corrosion-resistant materials in bulk, often unsuited (in bulk form) to applications requiring load-bearing or wear-resistant capabilities
Surface Engineering	• permits tailoring of surface properties offering enhanced corrosion resistance
	• offers high degree of flexibility and choice of most suitable material(s)
	• provides for optimisation in complex demanding engineering situations
	• offers cost-effective solutions to fulfil the design audit

two distinct headings: firstly, its role in reducing corrosion in aqueous environments at ambient temperatures (covering the field of activity often referred to as "wet corrosion") and, secondly, its adoption for the control of high-temperature degradation process ("dry corrosion").

15.2.1 CORROSION IN AQUEOUS ENVIRONMENTS AND ITS CONTROL BY SURFACE ENGINEERING TECHNOLOGY

Before examining this topic in some detail, there is a need to consider some key questions, as cited in Table 15.5. Obviously, in choosing, for example, a metal coating to protect a metal substrate, there seems to be a paradox: the coating clearly has to corrode more slowly than the underlying substrate. However, not only does the susceptibility of a given metal to corrosion need to be considered, but also the nature of the attack; for example, a tendency to pitting rather than more uniform corrosion has to be recognized. Some key aspects of corrosion and common forms of corrosive attack are identified in Table 15.6. In considering how surface engineering may reduce corrosion, the total range of possible methods, including this approach, needs to be considered (see Table 15.7), where it is seen that the use of coatings is merely one, albeit important, strategy. Notice in this table how other aspects include selection of materials and design, factors of major significance in surface engineering in general, as well as in the particular context of effecting corrosion control.

In determining the use and likely success of a protective coating or surface treatment to inhibit corrosion, some key aspects of corrosion need to be borne in mind (Table 15.6). Again, it is emphasized that the forms of corrosion observed with unprotected bulk metals (Table 15.6) may also occur with certain metal coatings systems; therefore, a knowledge of the particular susceptibilities properties of coatings is required at the *design stage,* and careful selection of a coating system is required. Again the concept of "fitness for purpose" has to be emphasized.

The nature of coatings (Table 15.8) also has direct implications for the corrosion performance of a coated component. Thus, both physical and chemical coating properties are relevant. A heterogeneous porous coating is undesirable, being instrinsically susceptible to corrosion, as well as ob-

TABLE 15.5. Surface Engineering and Corrosion Control: Some Key Questions.

1. Why do metals corrode?
2. How fast do metals corrode?
3. What forms/types of corrosion are observed?
4. How may corrosion control be effected?
5. How can surface engineering practices be utilised to control corrosion?

TABLE 15.6. Corrosion: Some Key Aspects.

- Electrochemical nature of attack
- Origins and types of galvanic cell:
  —dissimilar details
  —metal heterogeneities
  —surface films
  —heterogeneities in electrolyte
- Influence of water composition:
  —pH value
  —dissolved salts
  —dissolved gases
- Influence of flow, temperature, and heat transfer

Forms of Corrosion

- Uniform attack
- Galvanic or two metal corrosion
- Crevice or deposit corrosion
- Pitting corrosion
- Intergranular corrosion
- Selective leaching
  —dezincification
  —graphitisation
- Erosion corrosion
- Stress corrosion

TABLE 15.7. Methods for Corrosion Control.

- Selection of materials
- Design
- Contact with other materials
- Mechanical factors
- Coatings
- Environment
- Interfacial potential

TABLE 15.8. On the Nature of Coatings and Implications
for Corrosion Performance.

- Physical properties
  —density (porosity)
  —structure
  —morphology
- Chemical properties
  —thermodynamic considerations
  —prediction as linked to likelihood of reaction
  —No anticipation of speed of reaction: (protective oxide? passivity?)
- (Mechanical properties)

TABLE 15.9. Some Factors Influencing the Corrosion Performance of Coatings Systems.

- Adherence of coating to substrate
- Coherency in system–coating/substrate compatibility
- Porosity in coating: open or closed
  —gas access?
  —galvanic effects?
- Interdiffusion (in HT systems)

viously not being able to afford total surface protection: worse, pitting may encourage corrosion of the substrate. Information is required as to the thermodynamic stability of the coating in the environment; however, such data will give no indication as to speed of reaction. The coating may initially corrode, only to subsequently passivate and hence afford protection.

Other important factors influencing the corrosion performance of a coating system are set out in Table 15.9. It is to be noted that porosity in the coating may be open or closed: in the former case, access of the environment to the substrate becoming possible, setting up galvanic effects (which may be harmful or beneficial) or allowing access of gas in a high-temperature degradation situation (discussed later). At high temperatures, interdiffusion effects between the substrate and the coatings are also often life-limiting. Table 15.10 lists the "ideal" coating(s) system, something that is rarely totally achieved in reality.

Gabe [3] has pointed out that, although the choice of metal to be used as a coating may encompass a large number of elements in principle, in practice, the number is severely limited by cost considerations alone. Cheaper metals tend to be used as substrates or bulk constructional materials; the most expensive are used strictly as coatings, whereas intermediate metals such as nickel may be used as both, depending upon precise economic budget and technical requirements.

The factor of economics also strongly dictates [3] the "appropriate" thicknesses of coatings for a required service life. Thus, the life of a given coating thickness for a given metal is, in general, known and a decision taken as to optimum thickness for planned component lifetime. For example, the rate of atmospheric corrosion of Al in an industrial environment

TABLE 15.10. The Ideal Coating/Coating System.

- Thin, adherent, coherent (matched physical/mechanical properties)
- Chemically inert or slow rate of degradation (passivation?)
- Low porosity
- Good galvanic compatability w.r.t. substrate (if porous)
- Good interdiffusional stability w.r.t. substrate

is much slower than either Zn or Cd. Again, in general, a coating of sufficient (chosen) thickness will have an acceptable level of porosity to give planned lifetime at a chosen overall cost.

Earlier (Table 15.3), the historical dimension to surface engineering was cited, and in this context the developments in the use of Zn and Ni coatings are noteworthy. Zinc may be deposited on iron and steels, using a variety of techniques, and the well-established practices of hot-dip galvanising, sherardising, and electroplating have all been and continue to be used where process selection again relates directly to the coatings "fitness for purpose." It is interesting to note that, despite the successful uses of these coatings over many tens of years, it is a technology that has been continuously improved and adapted. The development of zinc-rich alloy electrodeposited coatings containing nickel or cobalt, for example, for improved corrosion resistance is a recent example in this pattern of continual evolution and improvement in the technology.

Thus, for example, some preliminary results [4] of the corrosion behaviour of Zn-Ni electrodeposited coatings with nickel contents in the range of 0.42–21.3% have been described. Here it has been shown in polarisation tests in aqueous sodium chloride that behaviour is critically dependent on alloy content, optimum performance in terms of corrosion rates (A/cm²) being observed with alloys of intermediate Ni content. Coatings with higher or lower Ni levels gave somewhat poorer performances, but in general, all of the Zn-Ni alloys exhibited better corrosion performance than unalloyed Zn. Such improvement in performance was mirrored in the measured corrosion potentials of the alloy coatings, increasing Ni content, shifting potentials to less negative values over the corrosion test period of some thirty-five days. Similar encouraging and interesting results were also reported for a series of Zn-Co alloy coatings (see Figures 15.1 and 15.2).

The potential for these coatings would seem to be in the thinner coatings, which could be used (on account of their longer life) and, hence, ease of fabrication and therefore reduced cost.

The use of nickel coatings to afford corrosion protection is an equally well known technology, where electrodeposition has traditionally been the favourite process. However, the invention in the late 1940s of the process of electroless deposition [5], allowing the production of thin, dense, pore-free, alloy coatings based on nickel, but containing phosphorus [6,7] or boron (8), heralded the arrival of a system capable of giving enhanced wear and corrosion performance especially for small, delicate, engineering components. Again, electroless nickel coatings have been the subject of development and enhanced sophistication. The incorporation of PTFE, for example, gives greatly improved frictional characteristics. Being heat-treatable, there is the potential for "tailoring" the costing to particular sur-

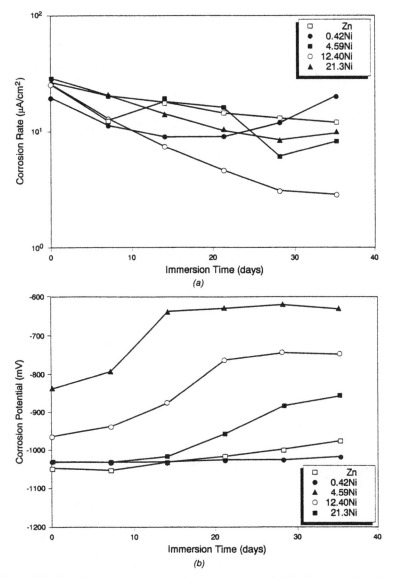

Figure 15.1 Corrosion rates and potentials for zinc and various Zn-Ni alloy coatings in sodium chloride solution [4].

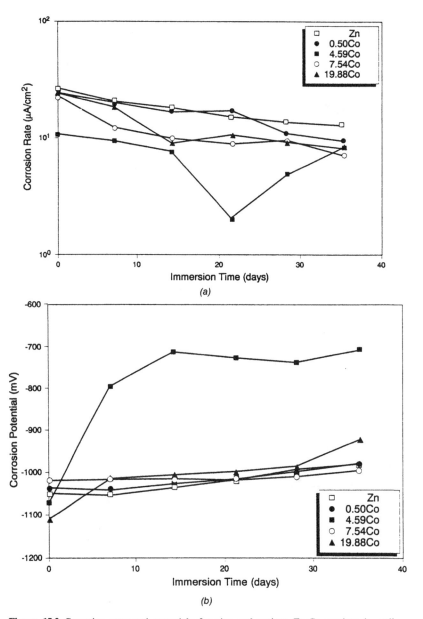

Figure 15.2 Corrosion rates and potentials for zinc and various Zn-Co coatings in sodium chloride solution [4].

face requirements—offering the best combination of wear and corrosion properties.

It is interesting to note that the alloy coatings produced by the electroless process and, indeed, by pulse electroplating are essentially amorphous and provide considerable potential for the tailoring/design of coatings for specific uses. Much research, however, is needed in this area. The potential of the electroless deposition process continues to attract the entrepreneur—for example, the use of electroless cobalt coatings for use in electonic applications.

In briefly reviewing past experiences and considering the future of electroless nickel, it is convenient to consider the process and material as a rival to electrolytic nickel. The advantages of electroless nickel lie in the almost perfect throwing power and in the dense, pore-free, coatings produced even in thin coatings. As a result, better corrosion resistance is exhibited for electroless nickel coatings (which are galvanically noble) on steel (active) substrates compared with electrolytic nickel for the same coating thickness. Such excellent corrosion performance, however, is only manifest with the nickel phosphorus variant: electroless nickel coatings containing boron at least at a low level of 0.3% show inferior corrosion resistance relative to both the nickel phosphorus variant and traditional electrolytic nickel coatings of the same thickness. It is clear that such coatings are porous and the leakage of solution promotes rusting of the steel substrate by galvanic action. Such porosity is caused by co-evolution of hydrogen during the overall cathodic reduction process, which prevents uniform nucleation and growth in these nickel-boron coatings [5,7,8].

An additional source of interest of electroless nickel-phosphorus coatings is a response to heat treatment [6]. By heat treatment, it is possible to substantially increase the hardness of such coatings, which, in turn, give very good resistance. Fully heat-treated nickel-phosphorus coatings give a wear behaviour that is not far short of the excellent performance traditionally associated with hard chromium coatings. It is believed that the superior wear behaviour of these coatings after heat treatment may be associated with higher fracture toughness with a retention of reasonable strength via precipitation coarsening. Nickel-boron coatings of higher boron content are also amenable to heat treatment, and the potential for these newer types of coatings has yet to be fully realised [8]. In principle, it should be possible to optimise both corrosion and wear performance by an appropriate coating selection and heat treatment processes. The matter is discussed and illustrated further in Section 15.3.

15.2.2 THE DESIGN OF COATINGS FOR THE INHIBITION OF HIGH-TEMPERATURE DEGRADATION

As noted in Table 15.11, the concern for high-temperature (H.T.) corro-

TABLE 15.11. **High-Temperature Corrosion: Forms/Modes of Attack.**

The concern for H.T. corrosion is its influence on specific component life by reducing the load-bearing cross section and introducing sources of stress concentration, via:

Surface Scaling	direct conversion of metal to corrosion product, decreasing sectional area and thus load-bearing capacity
Internal Degradation	further reduces load-bearing capacity and imposes stress concentration, which reduces fatigue resistance
Surface Scale Spallation	occurs as a result of heterogeneous scale growth (scale/metal mismatch) and exfoliation, especially during thermal cycling leading to subsequent enhanced corrosion
Corrosion Product Vapourisation	leading to loss of protective oxides, e.g., Cr_2O_3, reducing cross-sectional area, depleting the alloy of Cr and promoting higher oxidation rates

sion is its influence on specific component life by reducing the load-bearing cross section and introducing sources of concentration by four distinct processes as cited. It is to be emphasised that five types of H.T. corrosion processes are recognised (Table 15.12), including oxidation. In particular, H.T. coating and alloy design has, in general, been with reference to the development of oxidation resistance, and research concerning inhibition of these other wastage processes has received less systematic attention.

Strafford, Datta, and Gray [9] have discussed the design of coatings for such environments, pointing out it is necessary first to consider the various mechanisms of coatings degradation. These are indicated schematically in Figure 15.3. Coating degradation and ultimate failure take place by two processes, both of which rob or denude the coating of the elements present, by design, to effect resistance to the high-temperature corrosion process.

The prevention of failure by corrosion, i.e., mode (A), can, in theory, be achieved by adding to the basic metal M an element such as chromium to form an MCr alloy; chromium, for example, in an appropriate amount (about 20% in a nickel-chromium type material), would provide acceptable oxidation performance in many situations. However, the oxidation process obviously consumes chromium to form the protective oxide

TABLE 15.12. **High-Temperature Corrosion Processes.**

• Oxidation	• Carburisation
• Sulphidation	• Chloridation
• Hot corrosion	

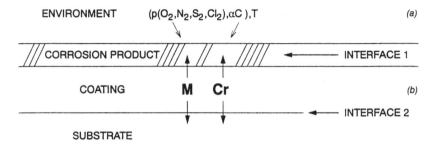

Figure 15.3 High-temperature coating degradation mechanisms: (a) through reaction with the environment at interface 1; (b) through diffusional interaction between coating and substrate at interface 2.

Cr_2O_3, and unless there is a sufficient reservoir of chromium in the coating, the concentration at interface 1 will be reduced by outward diffusion to such a level that there is insufficient chromium to prevent nickel oxide formation and rapid oxidation.

At the same time, however, the level of chromium in the coating is reduced through interdiffusional interaction between the coating and substrate — degradation model (B) — involving an inward diffusion of chromium and/or outward diffusion of elements constituting the alloy substrate (see Figure 15.3).

In the design of a coating there is thus a need to choose the right type and concentration of element (such as chromium) to reduce corrosion, and at the same time ensure that the mode (B) degradation is controlled. Often, this means the need for additional elements (which may assist in the promotion of corrosion resistance) or even an additional barrier layer between the coating proper and the substrate. It is believed that the flexibility of the physical vapour deposition process can be exploited here. The significance of composition in the design of a coating material to resist environmental degradation is illustrated in Figure 15.4.

In a coating of general type MCrAlX, the nature of M (normally iron, nickel, or cobalt or a mixture of these elements), the levels of chromium and aluminium, and the element X as a minor or major addition all need to be considered for optimum resistance to a particular environment. The development of a material with oxidation resistance is well understood and documented and will not be dwelt on here: Strafford [10] has reviewed this matter. In simple terms, resistance is associated with the presence of adequate levels of chromium (about 20–25%), aluminium 3.5% or 6–7%, and X as yttrium, zirconium, or hafnium at about the 0.15–1% level.

Recent studies conducted by Strafford et al. [11–13] have indicated the importance of the choice of the material base M in the development of re-

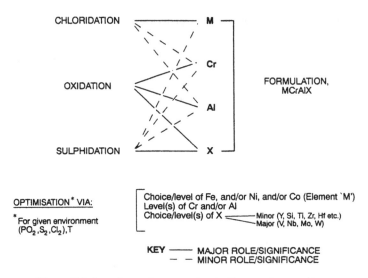

Figure 15.4 Corrosion performance: the significance of composition.

sistance to chloridation. Pure iron is readily attacked by chlorine gas to produce an iron chloride volatile corrosion product. Nickel, on the other hand, is relatively unaffected, and, by adding nickel to iron, an alloy base is produced with resistance to chloridation. The composition of alloy base is therefore a major factor in the design of a coating material to resist chloridation. The roles of chromium and/or aluminium additions are not apparently as important, although still significant. The significance of alloying element type and concentrations in the development of chloridation resistance has been discussed elsewhere in detail [9,13]. The chloridation behaviour of a number of metals and binary model alloys is illustrated in Figures 15.5 and 15.6.

If attention is now turned to the inhibition of sulphidation, the approach to coating design is rather different. Attention here focusses on the role of element X in an MCrAlX-type system, where X is a major, not minor, alloying addition, i.e., at a level of several weight percent. Sulphidation corrosion is a much more severe and rapid process than oxidation: Strafford and Datta [14] have recently reviewed this phenomenon and the problems in overcoming its aggressive nature. A major difficulty arises because additions to chromium, for example, nickel, are not particularly beneficial in inhibiting sulphidation. The reason for this observation is the fact that chromium itself sulphidizes many orders of magnitude faster than it oxidizes. As a result, conventional high-temperature coatings are of very limited use against sulphidation attack, and a new approach to coating

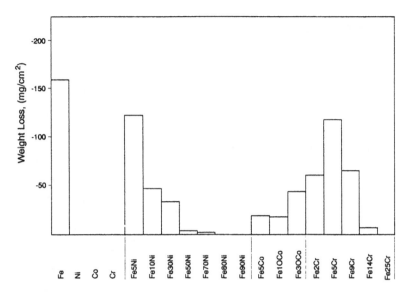

Figure 15.5 Observed weight losses per unit area for selected metals and alloys after exposure to the oxy-chlorine atmosphere ($pCl_2 = 10^{-8}$ atm, $pO_2 = 1.05 \times 10^{-16}$ atm) for 2 hr at 1000°C (Strafford et al.; courtesy Ellis Horwood Ltd.).

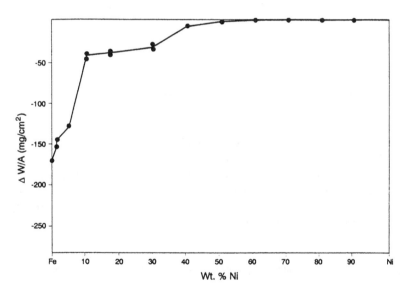

Figure 15.6 The influence of nickel content in iron-nickel alloys on the weight losses experienced by these alloys when exposed to an oxy-chlorine environment at 1000°C for 2 hr ($pO_2 = 10^{-16}$ atm, $pCl_2 = 10^{-8}$ atm) (Strafford, Datta, Forster; courtesy Ellis Horwood Ltd.).

257

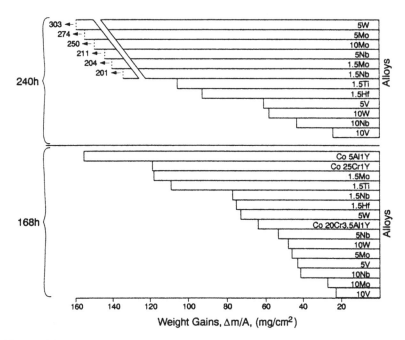

Figure 15.7 The influence of semi-refractory or refractory element type and content on the sulphidation behaviour of a Co20Cr3.5AllY control alloy after exposure to an H_2/H_2S environment ($pS_2 = 10^{-6}$ atm) at 750°C for 168 or 240 hours (other than the Co25CrlY material, all of the alloys contained the indicated type of elements within the five component control alloy) (Strafford, Dattta, et al.; courtesy Pergamon Press).

design is required. Strafford and coworkers [15] have shown, as predicted [16], that the Group V-VI refractory elements, for example, vanadium, niobium, and molybdenum, exibit very low rates of sulphidation, and this favours their use as elements X in MCrAlX-type materials. Figure 15.7 usefully summarises the beneficial influence of such refractory elements [17]. Such elements are also of interest in the context of the inhibition of secondary diffusion, i.e., the development of a barrier layer noted above. Here the refractory metals W, Nb, and Mo have good potential, with Fe, Co, and Pt being least efficient.

15.3 SURFACE ENGINEERING AND THE INHIBITION OF WEAR

Some of the features of wear and various wear processes are identified in Tables 15.13 and 15.14. It is to be noted that wear is probably the most common cause of direct or indirect (through fatigue) engineering component failure. As with corrosion, many estimates of the undoubtedly enormous cost−direct and indirect−of wear problems have been made.

TABLE 15.13. Surface Engineering and Wear.

Wear	• probably the most common cause of direct or indirect (e.g., fatigue) engineering component failure
	• expensive—e.g., savings of ~1.3–1.6% of UK GNP could be effected by attention to tribological problems
	• embraces many complex interactive forms-adhesive, abrasive, chemical, fretting, erosive
	• is a surface/subsurface phenomenon
Surface Engineering	• offers the choice and flexibility (design and materials selection) to effect optimisation remedial action specific to the problem

Surface engineering provides a means to reduce the problem, and, in principle, it should be possible to tailor coatings and/or surface treatments to effectively control or even eliminate wear. In practice, and in detail, the mechanisms [2] of wear are many and complex (Table 15.14), and as a result, the concomitant refinements in design and materials selection have yet to be achieved. Much more systematic research is needed before it may become possible fully to prescribe surfaces and treatments for the most effective and efficient inhibition of wear.

Figure 15.8 [18] summarises the various thicknesses of wear-resistant coatings that are in use and the associated process technologies. The range of coating thicknesses over seven orders of magnitude is, in itself, indicative of the many wear processes experienced, and the responses required for control and these data borne in mind with the recognised wear process cited in Table 15.14 underline and justify the observation that no single process of coating/surface treatment can be expected to cope with

TABLE 15.14. Wear Processes.

Abrasive	Removal of material from (and deformation of) a relatively soft surface when (relatively) hard material contacts (slides or rolls across) it.
Adhesive	Removal of material from a surface by welding together and subsequent shearing of minute areas that slide (or roll) across each other under pressure. Sometimes material transfer from one contact phase to another.
Chemical	Mechanical wear damage in presence of significant chemical reaction(s), e.g., acid- or sulphur-containing fluids in contact with mechanical bearings (or similar moving contacting surfaces).
Fretting	Wear mechanism between tight-fitting surfaces subjected to small, relative oscillatory motion of extremely small amplitude. Normally accompanied by corrosion, especially of the very fine wear debris—fretting corrosion.
Erosive	Removal of material due to impact of solid particles.

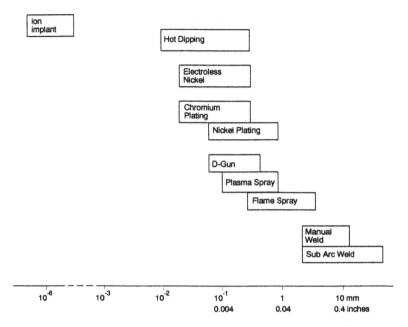

Figure 15.8 Typical thickness ranges of wear-resistant surfacings (after P. Furnival; courtesy Ellis Horwood Ltd.).

variables of these magnitudes. Again, then, as noted earlier, there is the need to recognise (and hopefully understand) a particular wear situation and design a surface engineering treatment/coating to contain the particular problem—"fitness for purpose."

Despite the complexity of the various overlapping recognised wear processes, progress is being made, through research and development, towards a rationale for the design of wear-resistant coatings. Here, the development of composite or hybrid, multilayer coatings, perhaps epitomized in the novel advanced coatings for enhanced cutting tool efficiency produced via chemical vapour deposition (CVD) (see, for example, Trent [19]), provides a glimpse of likely future success.

A useful example is provided in the continuing development and refinement of electroless nickel coatings, already cited in Section 15.2, as offering good corrosion resistance. Here, the prospect is to develop coatings not only offering adequate corrosion resistance, but also wear performance. Some of the advantages and limitations in the use of electroless nickel coatings are cited in Table 15.15, where it is noted that coatings can be hardened through heat treatment. Since high hardness usually is synonymous with improved wear behaviour, there is the prospect of achieving acceptable wear performance with adequate corrosion resistance. Two variants of electroless nickel coatings—those bearing phosphorus or boron, each at different concentration levels, depending upon both type and process variables—are known. The response of these coating types to heat treatment has been researched by Datta and Strafford et al., and the results are summarised in Figures 15.9 and 15.10. Ni-B coatings of low boron content (<1%) (Niklad 752) do not harden on heat treatment, while those with B contents of ~3% (Niklad 740) exhibit significant hardening, critically dependent in temperature, and peak hardnesses of ~1000 VPN being observed after 1 hr treatment at 200°C.

The abrasive wear behaviour of various N-P and Ni-B coatings is demonstrated in Figure 15.11, performances compared under the same wear conditions with mild steel and hard Cr plating (a traditional wear-resistant coating). Wear behaviour is clearly related to heat treatment condition, as well as coating type. The good performances of electroless Ni-B as-deposited and Ni-P (8%P) after heat treatment at 600°C (cf., Figure 15.10) are noteworthy.

Figure 15.12 details the abrasive wear behaviour of the Ni-high boron (Niklad 740) coating where, again, the important influence of heat treatments is evident, optimum wear behaviour being observed in a coating heat-treated (to maximum hardness, Figure 15.10 at 200°C), overaged, softened material (600°C) exhibiting poor behaviour, worse than that for the as-deposited coating. Figure 15.13 demonstrates the adhesive wear behaviour of a Ni-P (8%P) coating after various heat treatments.

TABLE 15.15. **Electroless Nickel Coatings.**

• Coating deposited by autocatalytic chemical reduction (electroless) of nickel ions by hypophosphite, aminoborance or borohydride compounds.
• Ni-P or Ni-B

Advantages
• Good coating/substrate adhesion
• Good corrosion and wear resistance
• Uniform thickness, good throwing power
• Solderability and brazability
• Low labour costs
• Can be hardened through heat treatment

Limitations
• High chemical costs
• Brittle
• Poor weldability of Ni-P deposits
• Slower plating rates
• Good substrate surface finish required

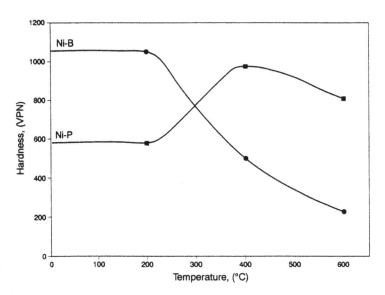

Figure 15.9 The hardness of nickel-phosphorus and nickel-boron electroless coatings following heat treatment for 1 hr at the indicated temperatures. The phosphorus content of the nickel-phosporus deposit was approximately 8%, while the boron content in the nickel-boron deposit was approximately 0.25% (Datta, Strafford, et al.; courtesy of Ellis Horwood Ltd.).

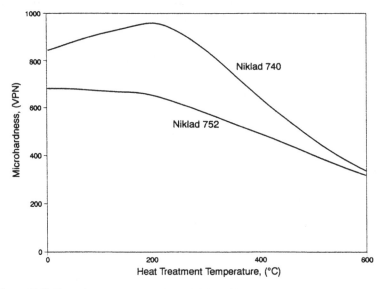

Figure 15.10 The variation in microhardness of 25-μm thick electroless Ni-B deposits under a load of 1 g (Datta, Strafford, et al.; courtesy Ellis Horwood Ltd.).

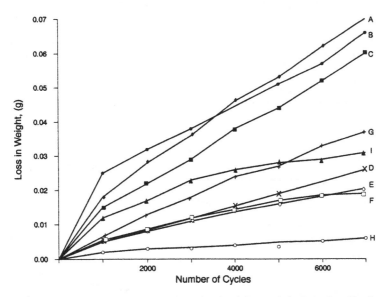

Figure 15.11 Effects of surface treatments on the abrasive wear behaviour plotted as loss in weight against number of test cycles: (A) mild steel (substrate material); (B) electroless Ni-P as-plated; (C) electroless Ni-P heat treated at 200°C for 1 hr; (D) electroless Ni-P heat treated at 400°C for 1 hr; (E) electroless Ni-P heat treated at 600°C for 1 hr; (F) electroless Ni-B as-plated; (G) electroless Ni-B heat treated at 600°C for 1 hr; (H) hard Cr as-plated; (I) carbonitrated (Datta, Strafford, et al.; courtesy Ellis Horwood Ltd.).

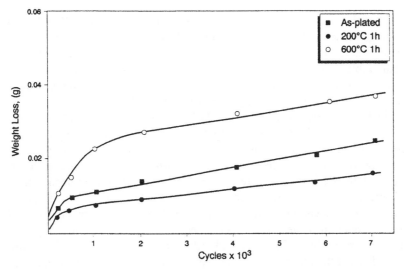

Figure 15.12 The abrasive wear behaviour of 40-μm thick high boron (Niklad 740) deposits under an applied load of 1 kg (Datta, Strafford, et al.; courtesy Ellis Horwood Ltd.).

263

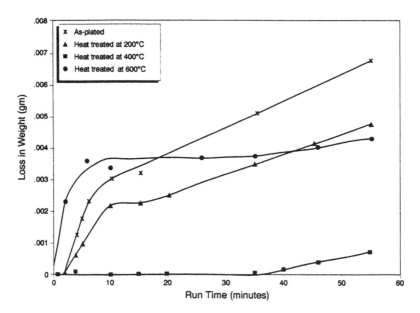

Figure 15.13 Adhesive wear behaviour of electroless Ni-P deposits plotted as weight loss against sliding time: × as-plated; ▲ heat treated at 200°C; ■ heat treated at 400°C; ● heat treated at 600°C (Datta, Strafford, et al.; courtesy Ellis Horwood Ltd.).

All of these data concerning the property of the various Ni-P and Ni-B coatings illustrate the potential for the "tailoring" of such coatings to meet requirements, including corrosion resistance. Further, in the context of wear, Table 15.16 identifies the benefits in situations where normal lubrication is difficult in the use of electroless nickel PTFE composite coatings; the limitations of these coatings are also indicated.

Finally, it is pertinent to recall the generally beneficial effects many coatings have on the fatigue life of engineering components. Again, however, caution is needed in the choice of coatings. Datta, Strafford, et al. have shown that, while electroless nickel coatings have a less damaging effect on fatigue life than a corresponding thickness of electroplated nickel, all such coatings reduced fatigue life (Figure 15.14). Figure 15.15, as a matter of detail, indicates the variation in fatigue life of 40-micron Ni-P coated steel in relation to heat treatment, benefits being shown following heat treatment procedures. This figure also demonstrates how fatigue life is improved as the thickness of as-deposited electroless nickel is reduced: a 15-μm coated sample giving the same fatigue performance as an uncoated sample.

15.4 CONCLUSIONS

It has been pointed out and demonstrated that, although surface engineering is now widely recognised as an enabling technology of special, strategic importance, it is, in fact, a very old technology. Thus, for example, while many surface-hardening processes such as carburising are of very ancient origin, e.g., in the attainment of keen cutting edges on steel tools and weapons, surface engineering in its many modern guises, such as physical vapour deposition (ion plating), is indeed, in the absolute forefront of technological advance.

The flexibility and multiplicity of surface engineering process technologies have been emphasised, features that are both advantageous and disadvantageous. In this latter context, too often, surface engineering has been seen and used as a "last resort" to provide an immediate solution ("often as a holding operation") to an unforeseen problem, often an engineering component service problem. In this chapter the need to focus on surface engineering as an essential element in the total design process/audit, ab initio, has been stressed. Here again, the need to consider the most suitable

TABLE 15.16.

Not applicable, where in absence of lubricant, seizure or galling can take place:

Electroless Nickel/PTFE

Added benefits of:

- Lower coefficient of friction
- Increased lubricity (dry lubrication)
- Plating method and prices very similar to electroless nickel

Limitations

- Soft particles at low concentrations may act as crack initiation sites
- Reduced fracture toughness
- Increased wear rates

  (easy fracture of adhered junctions on coating side of the sliding interface)

  ∴ High particle content to provide adequate surface coverage—complete barrier layer

Applications

- Ball/butterfly valves (oil/gas)
- Precision instrument parts
- Carburetor choke shafts
- Pump components

coating *system*, i.e., coating/surface treatment *and* substrate, has been emphasised.

Undoubtedly, one of the greatest strengths of surface engineering lies in its flexibility in terms of coating/surface treatment type and associated process technologies. These features offer real prospects for the "tailoring" of coatings/treatments for very specific purposes—to enhance corrosion or wear resistance for very particular circumstances, in conjunction with the most appropriate substrate condition.

Particular promise, it is believed, is held in the systematic development of multilayered coating systems where the specific requirements at the working component external surfaces can be entertained while ensuring compatibility at the crucial coating/substrate interface. In the long term, it should be possible to develop expert coatings/materials selection systems. To this end, however, much more sytematic research is required, both pure and applied, in order to have a much better understanding of the fundamental basis of the properties of bulk materials to be used in coatings, the properties of actual coatings, and the manner in which these can be con-

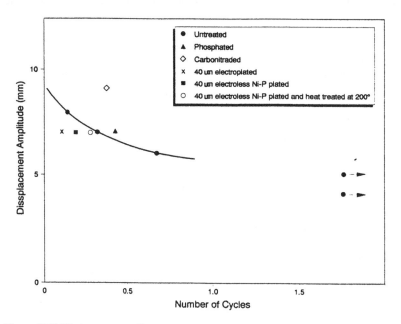

Figure 15.14 Displacement amplitude against the number of cycles to failure: ● untreated; ▲ phosphated; ◇ carbonitrated; × 40-μm electroplated; ■ 40-μm electroless Ni-P plated; ○ 40-μm electroless Ni-P plated and heat treated at 200°C (1 hr). Note that the two untreated specimens run at 5- and 4-mm displacement were tested for 5.65 and 3.04 M cycles, respectively, and did not fail (Datta, Strafford, et al.; courtesy Ellis Horwood, Ltd.).

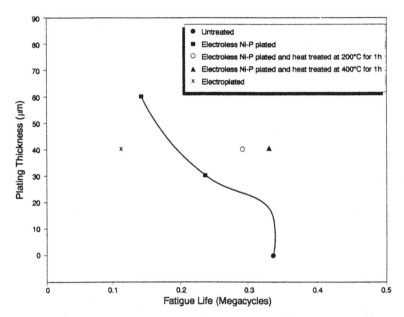

Figure 15.15 Comparison of fatigue life at various plating thicknesses: ● untreated; ■ electroless Ni-P plated; ○ electroless Ni-P plated and heat treated at 200°C for 1 hr; ▲ electroless Ni-P plated and heat treated at 400°C for 1 hr; × electroplated (Datta, Strafford, et al.; courtesy Ellis Horwood Ltd.).

trolled and tailored to meet specific requirements through the most appropriate process technologies.

15.5 REFERENCES

1 Strafford, K. N., Datta, P. K. and Gray, J. S. *Surface Engineering Practice.* Ellis Horwood, Chichester, UK, 1990.

2 Lansdown, A. R. and Price, A. L. *Materials to Resist Wear.* Pergamon Press, Oxford, UK, 1986.

3 Gabe, D. R. Chapter 4 in *Coatings and Surface Treatment for Corrosion and Wear Resistance*, K. N. Strafford, P. K. Datta and C. G. Coogan, eds., Ellis Horwood Ltd., Chichester, UK, 1984.

4 Abibsi, A., Short, N. and Dennis, K. "Electrochemical Investigation on Zinc Alloy Coatings," paper presented at UK Institute of Metal Finishing, Midlands Branch, Aston University, UK, 1988.

5 O'Donnell, A. "Properties and Engineering Applications of Electroless Nickel Coatings," M. Phil. CNAA Thesis, Newcastle-upon-Tyne Polytechnic, UK, 1982.

6 Strafford, K. N., Datta, P. K. and O'Donnell, A. *Mat. and Design,* 3:608, 1982.

7 Datta, P. K., Strafford, K. N., Storey, A. and O'Donnell, A. Chapter 3 in *Coatings and Surface Treatment for Corrosion and Wear Resistance*, K. N. Strafford, P. K. Datta and C. G. Coogan, eds., Ellis Horwood Ltd., Chichester, UK, 1984.

8 Datta, P. K., Strafford, K. N. and Allaway, S. Chapter 3.4.3 in *Surface Engineering Practice: Processes, Fundamentals and Applications in Corrosion and Wear*, K. N. Strafford, P. K. Datta and J. S. Gray, eds., Ellis Horwood Ltd., Chichester, UK, 1990.

9 Strafford, K. N., Datta, P. K. and Gray, J. S. Chapter 3.1.1 in *Surface Engineering Practice: Processes, Fundamentals and Applications in Corrosion and Wear*, K. N. Strafford, P. K. Datta and J. S. Gray, eds., Ellis Horwood Ltd., Chichester, UK, 1990.

10 Strafford, K. N. *High Temp. Tech.*, 1:307, 1983.

11 Strafford, K. N., Datta, P. K. and Forster, G. *Corr. Science*, 29:703, 1989.

12 Strafford, K. N., Datta, P. K. and Forster, G. Chapter 2.2.2 in *Surface Engineering Practice: Process Fundamentals and Applications in Corrosion and Wear*, K. N. Strafford, P. K. Datta and J. S. Gray, eds., Ellis Horwood Ltd., Chichester UK, 1990.

13 Forster, G. "Chloridation of Metals and Alloys," Ph.D., CNAA, Newcastle-upon-Tyne Polytechnic, UK, 1989.

14 Strafford, K. N. and Datta, P. K. *Mat. Sci. and Tech.*, 5:765, 1989.

15 Strafford, K. N. and Jenkinson, D. Chapter 26 in *Corrosion Resistant Materials for Coal Conversion Systems*, D. B. Meadowcroft and M. I. Manning, eds., Applied Science Publishers Ltd., London, UK, 1983.

16 Strafford, K. N. Paper 5/15 in *Institution of Metallurgists (London) Conference on Environmental Degradation of HT Materials*, Isle of Man, 1980.

17 Strafford, K. N., Datta, P. K., Chan, W. Y. and Hampton, A. F. *Corr. Science*, 29:775, 1989.

18 Furnival, D. Chapter 1 in *Coatings and Surface Treatment for Corrosion and Wear Resistance*, K. N. Strafford, P. K. Datta and C. G. Coogan, eds., Ellis Horwood, Chichester, UK, 1984.

19 See Trent, E. M. *Metals Cutting, 2nd Edition*. Butterworths, London, p. 154, 1989.

CHAPTER 16

Surface Degradation of Oxides, Ceramics, Glasses and Polymers

R. ST. C. SMART[1]
G. A. GEORGE[2]

16.1 INTRODUCTION

SURFACE degradation of materials used in engineering, caused by either chemical or physical action, constitutes one of the major costs to industry in pretreatment and maintenance. Selection of materials and quality assurance in materials supply have become major issues with considerably increased budget allocations over the last decade. The relatively new field of surface engineering has emerged in response to the requirements for control of surface degradation in increasingly demanding applications. Chemical environments have become generally more reactive with increases in pollution (e.g., acid rain) and higher technology in the design of engines, solid state devices, forming machines, high-throughput processes, etc. Use of particular materials in environments not previously experienced or in extended extremes of temperature, pH, and chemical reactivity is often implied. Equally, the increased demands of physical wear in abrasion, impact, tension, compression, etc., have led to research for materials, surface modifications, and surface coatings to meet these demands. The now well-known coatings based on the nitrides and carbides are already strongly associated with the "surface engineering" approach to wear and corrosion resistance. Safety considerations and predictable lifetimes are adding to the list of requirements placed on a material and its surface by the consumer and in many cases, legislation.

---

[1]Surface Technology Centre, University of South Australia, The Levels, SA 5095, Australia.
[2]Department of Chemistry, University of Queensland, Queensland 4072, Australia.

In view of these rapidly escalating demands, it is somewhat surprising to find that the basis of selection of materials for a particular requirement, surface preparation, or coating selection, and definition of the major factors in these decisions is not well documented and, in fact, appears to be rather haphazard, limited, and unsystematic. An assessment of the literature shows that the chemical and physical mechanisms causing degradation of the surfaces of engineering materials, in general, are poorly understood despite decades of intense research. Bulk properties of the materials, allowing assessment of major structural factors, are readily available and, in general, reliable for most materials. In contrast, it is usually not possible to obtain a clear assessment from a reference manual, textbook, or supplier of the kinetics of surface degradation of the material in particular environments or application. In practice, most surfaces are protected by coatings of variable integrity rather than by inhibitory treatments based on the mechanisms causing degradation of the material surface. Lack of information on these mechanisms is, at least in part, a primary reason for the use of costly surface coatings. It is obvious that a more detailed understanding of these mechanisms could provide remedies in surface (or bulk) modification of the material itself or in control of external conditions.

In this review, recent advances in the understanding of surface degradation mechanisms relating to oxides, oxide films on metal surfaces, ceramics, glasses, and polymers will be described. Much of this understanding has been made possible by the advent of new surface analytical techniques in photon, electron and ion spectroscopy, and microscopy. Combined with chemical analysis of the reactive medium in contact with the solid surface (i.e., solution, gas phase, other solid surface), it is often possible to define the major factors controlling the surface degradation. In the longer term, this approach may lead to more cost-effective ways of inhibiting surface degradation.

16.2 SURFACE DEGRADATION

The diversity of materials now used in engineering applications, to which surface engineering methods and techniques can be applied, is too large to encompass in a single chapter. The major part of this group is, however, made up of metals, their oxidised surfaces, bulk oxides, minerals, ceramics (single and multiphase), glasses, and polymers. The study and control of corrosion and wear in metals and their surface oxide layers is discussed in detail in two other chapters in this volume. This survey is limited to the bulk oxides. The principles applying to the study of this surface degradation can then be encompassed by the four groups: oxides, ceramics, glasses, and polymers. These general classes constitute materials that have a major literature of studies relating to both chemical and physical surface degradation. There are, of course, many overlaps between these groups as, for instance, oxide phases in ceramics, minerals, and glasses, as well as composites, e.g., glass-ceramics, multiphase (mineral) ceramics, and fibre-reinforced plastics.

The intention of this chapter is to survey the major mechanisms, with definition of rate-controlling factors, applying to surface degradation in each group.

The observed forms of surface degradation fall into two general categories, namely, apparent chemical (or electrochemical) and physical effects. In oxides, ceramics, and glasses, chemical degradation is generally called "corrosion," implying surface reactions, dissolution, and reprecipitation (e.g., phase changes). These effects are observed as pitting (due to dissolution), bleaching or discolouration (due to leaching), hydration, gel layer formation and cracking (due to reaction/drying/shrinkage), powdering (due to reprecipitation), and residue formation (due to surface impurity deposits). Physical effects, generally classed as "wear," implying loss of material, thinning, etc., include pitting and layering (formed during deposition), spallation (layer detachment), powdering (grain detachment), and cracking (due to lattice strain, defects, grain boundaries, and intergranular films). The presence of inclusions as impurities or aggregated particles also constitute physical defects leading to surface degradation. The particular case of devitrification (i.e., crystallisation) at high temperatures or in specific chemical conditions also occurs in glass surfaces. The observed forms of surface degradation in polymers constitute another group of surface effects. Chemical degradation is observed as blooming (e.g., whitening), discolouration (due to oxidation), and, in particular, yellowing (due to UV exposure). Physical effects are mostly cracking (due to crystallisation and changes in free volume), hardening and brittleness (due to oxidation, corsslinking, and crystallisation), and, in some cases, softening (due to chain scission and hydrolysis).

To avoid the use of costly paints and inert surface coatings, then, in order to place the control of these forms of surface degradation on a more systematic basis, it is necessary to define and understand the thermodynamic and kinetic factors controlling their surface reactions and wear. The hope for the future is that, with this understanding, it will be possible to design materials and inhibit their surface reactions so that surface coatings are not required.

Analysis of both the solid surface and the medium in contact with the surface is essential in order to derive the information required to determine chemical or physical mechanisms of surface degradation. Hence, in addition to bulk phase analysis, techniques for analysing the surface layers (i.e., <5 nm) and near-surface region (i.e., >0.5 μm) are required.

16.3 SURFACE ANALYTICAL TECHNIQUES

In the last years, a number of techniques for rapid analysis of the surfaces of solid material have emerged to join the better known, more established techniques of analytical scanning and transmission electron microscopy. The ability to analyse the first few atomic layers distinguishes these techniques from XRF, electron microprobe (EDX), and infrared spectroscopy, in which the analysis depth is usually >1 μm, i.e., "near-surface" region. There are a variety of techniques in this category. Table 16.1 summarises techniques for analysis of the surface (i.e., 0.4–5 nm) and near-surface (i.e., 0.5–10 μm) regions. The probe used to excite the surface can be photons (e.g., X-rays, IR, UV) , electrons, ions, or electric fields with the emitted, analysed beam similarly divided into photons, electrons, or ions. In this chapter, we will concentrate on the surface-specific techniques of X-ray photoelectron spectroscopy (XPS or ESCA), scanning Auger spectroscopy/microscopy (AES, SAM), and secondary ion mass spectroscopy (SIMS), together with the near-surface techniques of analytical scanning and transmission electron microscopy and Fourier transform infrared spectroscopy (FTIR) because these are more readily available for routine study of solid surfaces. In many cases, the information on the mechanisms of surface degradation is confined to the first few layers of atoms or ions. It is also well established that there are major differences in chemical composition, structure, and reactivity between the surface and bulk of material and that the surface characteristics very often determine the suitability, lifetime, and performance of the material and application. Failures are as often related to surface properties as to those of the bulk materials. Both design of the materials to meet particular specifications, such as chemical inertness, and the inhibition of surface reactions demand information on the structure and chemical form of the surface layers. When the analysis method samples up to 1 μm depth, information on surface atomic layers is generally lost in the noise background of the spectrum. The new surface-sensitive techniques are of paramount importance and have literally made possible the gathering of the information required for understanding.

The reasons for the extreme surface sensitivity of XPS, AES (SAM), and SIMS are essentially concerned with the mean free path of the emitted beam in the solid. Electrons and ions with energy 1000 eV only travel distances of 2–3 nm before losing energy by collision (e.g., interparticle, phonon, excitation, etc.) within the solid whereas photons of equivalent energy have mean free paths of the order of 1 μm. For XPS or AES (SAM), this limits the electrons escaping with unaltered kinetic energy to those emitted from energy levels in the first few atomic layers of the solid. Other electrons, escaping after inelastic collisions, contribute to a second-

TABLE 16.1. Summary of Surface Techniques (Analysed Beam).

Probe	Photons	Electrons	Ions
Photons	• Infrared spectroscopy (FTIR)* • Raman spectroscopy (LR)* • X-ray fluorescence (XRF)* • X-ray absorption fine Structure (XAFS)	• X-ray photoelectron spectroscopy (ESCA)** • UV-photoelectron spectroscopy (UPS)** • Photon-induced auger electrons**	• None
Electrons	• Electron microprobe (EDX)* • Appearance potential spectroscopy (APS)* • Cathodoluminescence* • Analytical electron microscopy (EDS, ATEM)*	• Auger electron spectroscopy (AES)** • Low-energy electron diffraction (LEED)** • Electron microscopy*,** • Electron-impact spectroscopy (EIS)**	• Electron-induced ion desorption (EIID)**
Ions	• Ion microprobe: X-ray (IMXA)*	• Ion neutralization spectroscopy (INS)* • Ion-induced auger electrons*	• Secondary ion mass spectroscopy (SIMS)** • Low-energy ion scattering spectroscopy** • High-energy ion scattering spectroscopy (RBS)*,** • Ion microprobe: ions (IMMA)*,**
Electric	• None	• Scanning tunnelling microscopy** • Field emission microscopy (FEM)**	• Field ion microscopy (FIM)** • Field desorption microscopy (FDM)**

*Near-surface > 0.5 μm.
**Surface < 5 nm.

273

ary electron background on which the photoelectron or Auger peaks are superimposed. In SIMS, the penetration depth is limited by the energy of the primary ions. In static SIMS, low accelerating voltages and low beam currents (e.g., $< 1 \times 10^{-7}$ A cm^{-2}) are used to limit the secondary ion emission to the top one or two monolayers. A full description of these three techniques will not be attempted here. Theory and experimental applications of XPS, SAM, and SIMS have been reviewed in detail in a number of books and journals, for example, References [1–6].

It is useful, however, to summarise the experimental basis of each of these techniques and the type of information available from them. XPS uses X-rays to excite electron emission from core and valence energy levels of atoms in the solid surface. The equation $h\upsilon = E_b + E_k + \phi$, where $h\upsilon$ is the X-ray energy, E_b is the binding energy of the electron in the particular energy level of the particular atom, E_k is the kinetic energy of the emitted electron, and ϕ is the work function of the spectrometer, allows calculation of the binding energy E_b when the kinetic energy of the emitted electron is analysed in the spectrometer. In general, there are shifts in the binding energies of particular atoms according to the oxidation state, chemical bonding, and compound type in which the atom is found. Thus, for instance, it is possible to distinguish between carbon as carbide, graphite, $-COC-$ linkages, $-C=O$ species, and carbonates. In most cases, the analysis area for XPS is relatively large, i.e., 2–10 mm, except for specialised small spot size instruments, i.e., 150–250 μm.

SAM uses a focussed electron beam with energies in the 5–50 keV range to cause ionisation in core levels of surface atoms. The Auger process is a multi-electron event in which the primary electron beam causes electron ejection from a core level, an electron from a higher energy level drops into the hole left by the first electron, and the excess energy is dissipated by ionisation of the second (Auger) electron from a higher energy level. The kinetic energy of the Auger electron is measured and is again specific to the energy levels of the particular atom in which the Auger event occurred. The advantage of the technique is that focus of the electron beam can be achieved down to 20 nm with scanning as in the electron microscope. Consequently, SAM combines the physical imaging of the surface with surface chemical analysis of very small areas. The disadvantage, compared with XPS, is that only very limited shifts or changes in the signal occur with differing chemical states of the element. In addition, with materials such as polymers, the focussed electron beam produces degradation of the material during analysis.

The SIMS technique uses a focussed ion beam, usually Ar$^+$, Cs$^+$, O$_2^+$, or O$^-$, to sputter atoms, molecular fragments, and ions from the surface layers of the solid. The secondary ions, either negative or positive, are detected using a high sensitivity mass spectrometer. The ion beam can be scanned for lateral distributions of particular elements or molecular species. In the static SIMS mode, the sputter rate can be limited to removal of the equivalent of a single monolayer over periods longer than 1 hr. The technique is particularly useful for depth profiling to give compositional variation through surface layers and thin films. Because of surface degradation in dynamic SIMS, static SIMS is the technique of choice for studying polymers. A summary of the information available from these three surface-specific techniques is given in Table 16.2.

In contrast, the three techniques most commonly used to analyse the near-surface region of solids detect emitted photons either as X-rays or infrared radiation. In the most accessible technique, analytical scanning electron microscopy, analysis of X-rays generated by the primary electron beam, is carried out using energy dispersive spectroscopy (EDS). SEM can normally identify surface features to a practical limit of 100 nm, but

TABLE 16.2. Surface Techniques—Characteristics of XPS, Auger Spectroscopy, Scanning Auger Microscopy, and SIMS.

X-ray photoelectron spectroscopy (XPS) provides

- Quantitative surface compositions (to $\sim \pm 10\%$) for all elements from He to U down to ~ 0.1 atomic% at depths of ~ 0.4–5 nm
- Chemcial states, i.e., bonding, valency, of elements in the surface, including proportions of the same element in different forms, e.g., oxide/metal, sulphide/sulphate, etc.
- Quantitative depth profiles for all elements from angular resolved (nondestructive) spectra and from ion etching
- Relatively large analysis area (250 μm–10 mm)

Auger spectroscopy and scanning auger microscopy provide

- Quantitative surface compositions (to $\sim \pm 20\%$) for elements He to U at depths ~ 0.4–3 nm
- Physical imaging of surfaces
- Spatial distribution and chemical mapping of elements to ~ 20 nm–0.5 μm resolution
- Spot analysis of regions and crystallites
- Compositional profiling
- Limited chemical state differentiation

Secondary ion mass spectroscopy (SIMS) provides

- High sensitivity detection of surface elements and trace impurities to ppm levels for all elements from hydrogen to uranium over large dynamic range (1–10⁷)
- Monolayer depths of analysis
- Rapid depth profiling for this films, surface coatings, surface layers
- Identification of grain boundary and fracture face segregants using static SIMS profiling
- "Fingerprint" identification of polymers and their surface reactions using static SIMS

on the newer field emission instruments, this has been extended to < 1 nm. In the backscattered electron imaging mode, the technique is relatively sensitive to the surface region (i.e., 50 nm–0.5 μm) and can give contrast based on the average atomic number of the region or phase examined. Hence, it can be used to differentiate grains in a multiphase ceramic or different materials in a composite. Topographic images, obtained by combining secondary and backscattered electron images, can also reveal detail of pits, protrusions, precipitates, and altered regions on the surface. With EDS analysis, chemical maps of the lateral distribution of elements in the near-surface region can be obtained. Transmission electron microscopy (TEM) and scanning transmission electron microscopy (STEM) require preparation of thinned sections of the samples and are not as readily applicable to routine characterisation, although new instrumentation has very significantly reduced the time for these preparations. When necessary, TEM and STEM can provide much higher resolution for images of surface features down to unit cell level giving detail on atomic steps, ledges, corners, defects and reaction sites, intergranular films, and interphase regions.

All of the above techniques are *ex situ* techniques requiring high vacuum or ultra high vacuum for study of the solid surface. FTIR, with improved sensitivity and speed of analysis, can provide information on the molecular structure of adsorbed and near-surface species without removal from the reactive environment causing the surface degradation, e.g., a solution, reactive gases. This technique is particularly valuable in providing evidence of reaction in surface layers and changes in chemical species when coupled with the *ex situ* measurements from XPS. The theoretical and experimental basis for its use is well documented [7,8]. Table 16.3 comprises the main characteristics of these three near-surface analytical techniques.

It should also be noted that there are now several excellent computer codes for theoretical simulation of speciation and precipitation based on data from solution analysis. The EQ3l6 code [9] has been used for this purpose in studies of surface degradation of ceramics and glasses.

This package comprises two main data-handling programs, EQ3NR and EQ6; a substantial data base; and subsidiary programs for amending this data base. EQ3NR calculates the distribution of the aqueous species in a system from input data, including solution composition, temperature, pressure, and redox conditions. The calculated model of the aqueous solution includes the concentrations and thermodynamic activities of all species occurring in the model, as long as they are included in the data base, and also the minerals with respect to which the solution becomes saturated. Part of the output from EQ3NR serves as a starting point for EQ6, which calculates a simulated reaction path for the system using one of several mass transfer models available in the code. Initially, EQ6 brings the system to equilibrium by simulating precipitation of any supersaturated

TABLE 16.3. Near-Surface Techniques—Characteristics of Analytical
SEM, TEM, and FTIR.

Analytical scanning electron microscopy (SEM/EDS) provides

- Secondary electron images of surface features to (normally) ~0.1 μm in solution
- Backscattered electron images of phase differentiation, precipitates, reaction regions, etc., based on average atomic number contrast
- Topographic images of pits, protusions, reacted regions, etc.
- Scanning maps of elemental distributions in near-surface region normally for elements above F, i.e., not for C, N, O
- Quantitative analysis, with standards, for comparison with assays, XPS, etc.

Analytical transmission electron microscopy (ATEM) provides

- High resolution (i.e., <0.2 nm) imaging of surface features, e.g., defects, facetting, steps, ledges, pits
- Analysis of small particles (i.e., <0.5 μm)
- Intergranular boundaries at films, specific attack
- Preferential dissolution or surface reaction, e.g., specific facetting, phase preference, boundaries, etc.
- Precipitation and recrystallisation *in situ*
- Phase identification (composition and strictira) in grans by EDS and selected area electron diffraction

Fourier transform infrared spectroscopy (FTIR) provides

- Molecular structural information on adsorbed species
- Evidence of reaction in surface layers and changes in chemical species
- *In situ* studies of adsorption and reaction on solid surfaces in solution, air, or gaseous atmospheres
- Rates and extents of hydrolysis in surface layers

mineral phases and recalculating the solution parameters to satisfy the original mass constraints. The system may be open and the precipitates removed (i.e., flow-through), or it may be closed with the precipitates in equilibrium with the solution. Reaction progress may be described by time, temperature, and pressure changes; irreversible reactions between the solution and specified reactants; or a combination of these variables. EQ3/6 calculates changes in solution composition, occurrence and stability range of precipitated phases, and the point at which the solution is saturated in respect to the reactants. The reactants can be any solid phase in the data base or a gaseous component, e.g., atmospheric oxygen.

The combination of solution analysis, giving rates of loss of specific elements to solution via the EQ3NR code, and the surface and near-surface techniques listed above have been shown to be able to define the major mechanisms applying to the dissolution of oxides, ceramics, and glasses [10–12]. This combination of techniques has still been used relatively rarely for study of polymers, but the surface analytical techniques have been widely applied.

16.4 MECHANISMS OF REACTION AND EXAMPLES

Most of the information on surface reaction of oxides, ceramics, and glasses come from studies in aqueous solutions at different pH. Weathering reactions in ambient conditions show essentially the same mechanisms, but the thin (molecular) layers of adsorbed water, in this case, result in high local pH and supersaturation so that some of the mechanisms are exaggerated compared with bulk solution attack. Most polymer studies have been concerned with the oxidation of hydrocarbon polymers by a free radical chain reaction leading to surface degradation, but hydrolysis of certain functional groups may also be a significant degradation reaction.

16.5 SURFACE DEGRADATION OF OXIDES

16.5.1 MECHANISMS

This section deals with the surface reaction from dissolution and surface degradation of simple binary oxides according to the present understanding of mechanisms of reaction at their surface. The more complex mineral oxides (e.g., silicates, titanates, aluminates, and oxide glasses) will be discussed below. It is important to note at the outset that the rates of dissolution of these oxides under the same conditions of pH and solution temperature can vary by more than ten orders of magnitude from the fast-dissolving ionic oxides like MgO to the almost insoluble oxides like TiO_2 [10]. With this range of dissolution rates, it is unlikely that the same rate-determining step applies to different oxide categories or even to oxides in the same category with very different dissolution rates. Different mechanisms determine the dissolution rate in fast, intermediate, and slow kinetic regimes. Physical effects on the dissolving surfaces, such as selective faceting, pitting, and structural reorganisation, also vary widely from the ionic to highly covalent oxides.

Table 16.4 summarises the categorisation of oxides, their relationship to dissolution rates, and the predominant dissolution processes in each category [10]. For ionic oxides, surface degradation is dominated by initial surface reconstruction in solution or ambient air [13,14], at low pH by protons transferred to the oxide ions [15], and, at high pH by diffusion kinetics of protons and precipitation of metal hydroxide species [16]. Hence, ionic oxides dissolve relatively rapidly and tend to show dissolution pitting and precipitated colloidal particles drying to fine powders. Hydration reactions are confined to the first few layers of the reacting surface. An illustration of surface restructuring of MgO immediately after exposure to solution is shown in Figure 16.1 [15].

TABLE 16.4.

Categorisation of Oxides and Relationship to Dissolution Processes

	OXIDES	
Ionic	Semiconducting	Covalent
(ions at surface)	(partial ionicity at surface)	(lattice network of covalent atoms)
↓	╱ ╲	↓
insulating	p-type n-type conductivity conductivity	insulating
↓	╱ ╲ ╱ ╲	↓
	high low high low	
↓	↓ ↓ ↓ ↓	↓

Oxide Dissolution Rates (Examples)

		↓	↓	↓	↓		
↓							↓
Very fast		Fast	Slow	Fast	Slow		Very slow
MgO		CoO	NiO	ZnO	$\alpha\text{-Fe}_2\text{O}_3$		$\alpha\text{-Al}_2\text{O}_3$
CaO		MnO	CuO	CdO	MnO_2		TiO_2
BeO		TiO	UO_2		SnO_2		SiO_2

DISSOLUTION PROCESSES

Ionic	Semiconducting	Covalent
Surface reconstruction	Charge transfer to surface	Base-catalysed hydrolysis*
↓	↓	↓
Alteration of surface charge	Ion formation (M^{n+} and O^{2-})	Ion formation (i.e., $M(OH)_x^{y+}$
↓	↓	↓
M^{n+} transfer to solution	M^{n+} transfer to solution	$M(OH)_x^{y+}$ transfer to solution
↓	↓	↓
H^+ transfer to O^{2-} ion	H^+ transfer to O^{2-} ion	H^+ transfer to oxide species
↓	↓	↓
OH^- (or H_2O) transfer to solution	OH^- (or H_2) transfer to solution	OH^- transfer to surface

*e.g., $<\!Al \overset{O}{\diagup\,\diagdown} Al\!< \; + OH^- \rightarrow \;>\!Al\!-\!OH + {}^-O\!-\!Al\!< \;\overset{(H_2O)}{\rightarrow}\; 2(>\!Al\!-\!OH) + OH^-$

TABLE 16.4. (continued).

Factors Influencing the Reactivity of Oxides in Aqueous Solution

ROLE OF SOLID OXIDE

Factors		Relationship to Reactivity
Degree of ionic or covalent character in bonding	:	Nature of surface reaction; ion formation in the surface; charge transfer; protonation
Nature and concentration of lattice defects	:	Charge transfer; semiconducting
Nature and concentration of alloy dopants	:	Charge transfer; surface charge
Morphology and surface facetting	:	Surface reconstruction; concentration of surface sites; protonation; surface exchange
Nature and concentration of active (defect) surface sites (e.g., valence states, coordination)	:	Ion formation; protonation; effect of redox potential
Specific features of surface states	:	Redox effects; adsorption; surface conductivity

ROLE OF SOLUTION

Concentration of reactants, products, ionic additives	:	Kinetic regime (initial, advanced, near-saturation); potential control
pH	:	Protonation, base-catalysed hydrolysis; potential control; diffusion
Redox potential	:	Ion transfer; concentration of active sites
Complexing agents	:	Ion transfer; adsorption; surface charge
Surface active agents	:	Surface blocking; ion transfer; surface charge
Temperature	:	Reaction-mechanism category; surface reconstruction (and phase changes)

Semiconducting oxides divide into two groups according to their semi-conductivity [10]. Those with high conductivity have dissolution and surface degradation reactions dominated by the same mechanisms as for ionic oxides. The slow-dissolving, semiconducting oxides, however, have rates determined by charge transfer to the surface in order to facilitate formation of ionic species or transfer to solution. Their reactions are relatively slow and are strongly dependent on pH and on the defect concentration in the oxide. This, in turn, depends on the conditions of preparation of the oxide, either as an oxide film on the metal surface, as a deposited layer, or as a bulk oxide material. For instance, the dissolution rate in acidic solution (pH 1–3) of nickel oxide prepared from the carbonate at 700°C is more than 4000 times faster than that prepared at 1450°C. With surface area differences removed, the increase is still more than 400 times (i.e., 410 and 1.54 mol m^{-2}, respectively), due entirely to differences in defect concentrations [17]. A summary of this dependence and measurements of reaction rates can be found in Reference [10]. Surface degradation is observed in these oxides in similar forms to those in the ionic oxides, i.e., total dissolution, dissolution pitting, and precipitation giving colloidal and powdered products.

The covalent oxides have reaction and dissolution rates dominated by the base-catalysed hydrolysis reaction (Table 16.4) in which a network of -MOH species are formed in a hydrated, usually amorphous, gel layer [10–12,15,18]. This particularly applies to silica and alumina surfaces, but, for titania and zirconia surfaces, the rates of attack are extremely slow (i.e., $<1 \times 10^{-15}$ mol m^{-2} s^{-1}) and there is little evidence for formation of any extensive gel layer. The gel layer can incorporate cations, e.g., Fe^{3+}, Co^{2+}, Mn^{3+}, causing discolouration. In addition, upon drying, these heavily hydrated layers can shrink and crack, in some cases detaching from the surface, exposing new, unreacted oxide's surface for continuing

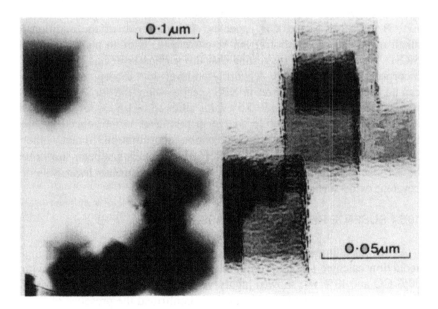

Figure 16.1 Transmission electron micrographs of MgO smoke as grown, and after <5 s in pH 3 HNO$_3$. Phase contrast reveals that the initially smooth cubes are roughened after immersion, but before dissolution commences [13].

reaction [18–20]. If the layer remains in place, it can inhibit further reaction by providing a diffusion barrier to reactants and products.

16.5.2 EXAMPLES

Three case studies will provide examples of impurity incorporation, surface reaction, and the effect of contaminant residues on oxide surfaces. they have been chosen to illustrate the application of the surface analytical techniques to these materials.

16.5.3 SURFACE IMPURITIES AND HYDROLYSIS

Silica, as a 500-nm thin film, is used as an insulating, protective layer on integrated circuit boards to isolate components. Intermittent electrical breakdown between the components has been attributed to contamination of the silica film surface, but standard analyses with SEM/EDS, XRF, and other near-surface techniques failed to show any evidence for contaminating species. XPS and AES analyses of the surface region also failed to detect signals additional to the silica species. Figure 16.2 shows SIMS spectra from a normal silica film (control specimen) and a film giving electrical breakdown in the insulating surface [21]. Comparison of the spectra shows that the contaminated specimens contain higher levels of Li^+, B^+, Na^+, K^+, and CH_x^+ species. The presence of the additional alkali metal cations and boron can be correlated, with an increase in the $SiOH^+$ species strongly suggesting that the surface layer has hydrolysed, incorporating these species. A hydrolysed layer with incorporated cations can be shown to be conductive in XPS spectra where the surface charging is reduced from values of 4.0–5.5 eV for silica to 0–1.5 eV for hyrolysed silica surfaces. Hence, mobile charge in this layer is introduced by the base-catalysed hydrolysis reaction and cation incorporation. In many other SIMS studies, the presence of the MOH^+ ion species has been shown to be a reliable indication of the extent and depth of surface hydrolysis of covalent oxide surfaces [11].

16.5.4 SURFACE REACTION AND PASSIVATION

A hydrometallurgical, reduced nickel/nickel oxide product, formed in a reduction calciner (Table 16.5) at 1400°C using fuel oil reduction with 10% CO and 10% H_2, showed highly variable rates of dissolution in acid for eletroplating baths [22]. XPS analysis confirmed the presence of Si as silicate in the surface of the nickel oxide particles and in the oxide surface layers on nickel metal particles separated magnetically. A comparison of surface composition between samples showing acceptable dissolution

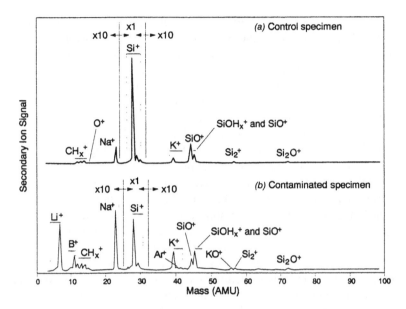

Figure 16.2 SIMS spectra of SiO₂ films from a control specimen and from a contaminated integrated circuit. (Reprinted with permission from Reference [21].)

rates, slow dissolution, and dust recovered from the calciner illustrates the level of Si incorporation into these surface layers (Table 16.5). XPS and FTIR were used to study the form of the incorporation by reacting vapour-deposited silica with bulk nickel oxide. The silica is incorporated by reaction into only the first few (i.e., three to five) atomic layers of the oxide surface, modifying the molecular structure to that of a nickel silicate thin film. Up to 12 at.% Si can be found in these layers by incorporation under reducing conditions. The surface specificity of the reaction is seen in Table 16.5 where the Si signal is halved by removal of 2.5 nm using Ar^+ ion

TABLE 16.5. XPS Surface Concentration (at.%) from Ni/NiO Products.

Dissolution Element	Acceptable		Slow		Calcine Dust	
	Initial*	25 Å Depth**	Initial*	25 Å Depth**	Initial*	25 Å Depth**
Ni	34.1	48.4	35.8	49.9	15.7	42.4
O	47.3	34.4	46.3	35.2	34.9	31.2
C	15.1	15.4	14.8	13.3	46.3	25.0
Al	0.0	0.1	0.0	0.2	0.0	0.0
Si	3.4	1.7	3.1	1.4	2.9	1.4

*Initial unmodified surface.
**Surface after ~25 Å etch using Ar^+ ion beam.

etching. The incorporation of silica into nickel oxide has the effect of lowering the dissolution rate of the oxide in acid at pH 2 by factors of more than 1000 times [23]. FTIR identified the vibrational structure of the SiO_4^{2-} groups in the surface layers [23]. In the table, the surface Si content appears to be similar between samples that were relatively easy to reduce and those that required extended reduction. Analytical TEM revealed, however, that the former sample was relatively uniform with particles of similar structure and composition, whilst the latter sample had extreme variation of Si/Ni ratios from 1.8 to 88% using EDS analysis. Figure 16.3 shows that the "poor" sample and the dust sample also have another characteristic, namely a higher proportion of reduced Ni metal due to more extended reduction required to overcome the poor dissolution properties. This is evident in the initial rapid dissolution in these samples. After adjustment for differences in surface area, it is clear that the long-term dissolution rates for these two samples indicate severe passivation due to the formation of silicate in the surface layers. Conversely, the very significant improvement in corrosion resistance due to reaction with silica is being exploited in current research at the University of South Australia to provide a thin film protection for metals via their oxide surface layers.

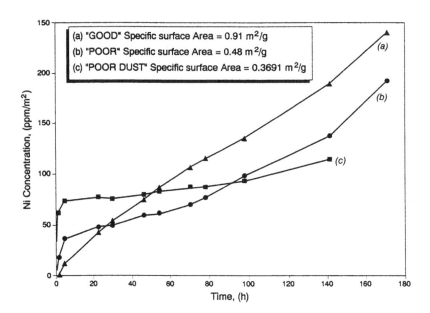

Figure 16.3 Dissolution kinetics of Ni/NiO product in 10^{-2} mol dm^{-3} nitric acid at 25°C. The dissolution rate (ppm m^{-2} h^{-1}) is given by the gradient of the curves at any particular time. (The "dust" is collected from bag filters in the exhaust gas system.)

16.5.5 SURFACE CONTAMINATION AND ADHESION

The chemical and physical forms of contaminant residues on surfaces, either as impurities or as a result of processing conditions (e.g., mould materials, co-deposited oxides), can be monitored by these techniques. For instance, XPS, SAM, and SIMS have been used to study the problem of ineffective detachment of deposited copper from stainless steel starter sheets in electrorefining [22]. In normal operation, detachment of the copper sheet is relatively easy, but, for contaminated stainless steel sheets, the copper layer can only be detached using jackhammers. Surface roughness, monitored using SEM and SAM, was similar for good and poor stainless steel, suggesting that the chemistry of the surface was more likely to be responsible for the differences in detachment. XPS revealed a carbonaceous layer containing copper and iron cations more that 10 μm thick on the surface of the adherent sheet. The carbon was partially oxidised giving, for instance, a composition at ~ 1-μm depth of 74% C, 23% O, 2.5% Cu, and 0.5% Fe. The oxidised carbon material was resistive and apparently polymeric in form, i.e., not graphitic, suggesting residues of humic/fulvic material acting as a "glue" at the interface. Cathode sheets showing relatively easy detachment also had some indication of the same type of carbonaceous layer, but this was much thinner and ion etching to a depth consistent with surface roughness entirely removed it. The concentration of Cu^{2+} in this layer was also much lower (i.e., <0.6%). The oxidised carbon layer apparently builds up on the surface during cycling of the sheets in cleaning, pickling, and pretreatment, whilst the oxidised metal species comes from corrosion of the stainless steel downgrain boundaries, as revealed by metallography of the underlying metal. SIMS revealed a range of other cationic impurities at trace levels in the carbon layers. Similar types of layers have been found adhering to membrane surfaces using a combination of surface analytical techniques and analytical electron microscopy.

Other examples of surface analysis of oxide surface layers and bulk materials have shown the results of surface treatments of steel pipe before fusion-bonded epoxy application, revealing thin phosphate layers with partial separation of chromate from silicate in the passivation treatment and aluminosilicate layers on glacial quartz used for glass making, resulting in severe handling and melting difficulites [24].

16.5.6 CONTROL OF OXIDE SURFACE REACTIONS

The surface reactions and corresponding physical alteration of binary oxides are relatively straightforward. Inhibition of these reactions can be achieved with control of (1) defect concentrations in the bulk oxide and its surface; (2) pH of the solution phase; (3) grain size of the oxide in poly-crystalline form; (4) integrity of oxide surface layers; (5) exclusion of par-ticulate impurities and aggregates; (6) exclusion of residues and deposits from other solid, liquid, or gaseous phases in contact with the surface. With the covalent oxides, it is also important to avoid cycles of solution (or wet ambient air) followed by drying, in order to inhibit surface hydroly-sis/shrinkage/cracking and/or spallation. The choice of oxides for particu-lar applications can be based on the oxide type and, for semiconducting oxides, the conductivity of the material. Ionic and semiconducting oxides can be made considerably more inert by selective surface reactions, e.g., with silica.

16.6 SURFACE DEGRADATION OF CERAMICS

16.6.1 MECHANISMS

In engineering applications, the ceramics can be divided into single-phase and multi-phase materials. The single-phase materials cover a wide range of compounds for electronic applications (e.g., perovskites, super-conductors), corrosion resistance (e.g., oxides as above, nitrides, bo-rides), hardness, and wear resistance (e.g., carbides, nitrides, borides). These single-phase ceramics can be used as bulk materials, often pressed or sintered to other materials (e.g., tungsten carbide), but also as thin films deposited, using PVD or CVD methods, on materials requiring protec-tion. Multi-phase ceramics are often used for incorporation of a range of waste elements in disposal technology (e.g., Synroc for high-level nuclear waste disposal).

16.6.2 PEROVSKITE DEGRADATION

The surface reaction and structural reorganisation of the perovskite $CaTiO_3$ have been studied in detail using the combination of techniques described in this chapter [11,15,25]. The compound will be used as a model to describe the major mechanisms controlling surface degradation of single-phase ceramics. Examples from other carbides, nitrides, and bo-rides will then be described. The essential features of single-phase ceramic dissolution are

- There is an initial, near-instantaneous release of ion-exchangeable cations either to solution (in aqueous attack) or to the exposed surface (in ambient weathering). This release arises from ions exposed by preparation procedures, particularly at defect (or kink)

sites. The amount released corresponds to a fraction of the monolayer [11].

- Ion-exchange reactions continue, but the depth of this exchange is limited to the first one to three atomic layers; i.e., "leaching" is limited to this depth [15]. Surface defects, intergranular films, exposed pores, and microcracks contribute to this ion exchange.

- Dissolution of the lattice of the crystalline ceramics occurs via the base-catalysed bydrolysis reaction (Table 16.4), disrupting the $M-O-M$ bonding and releasing further cations from this reacted layer. Evidence for this reaction is found in XPS and SIMS spectra where the hydrolysed $M(OH)_x^{y+}$ species are observed [25]. For perovskite, a limiting thickness, arising from retention of hydrolysed Ca^{2+} species in the layer, of ~ 10 nm is found. With the relatively high pH (>9) induced by the ion exchange and the retained Ca^{2+} ions, further reaction is inhibited because the system is driven into the thermodynamic stability field of the perovskite [26]. At higher temperatures, i.e., >90°C, however, the hydrolysed layer recrystallises as TiO_2 polymorphs (i.e., brookite, anatase) on the perovskite substrate as powdered products [27]. The kinetics of the hydrolysis reaction are also affected by local surface variables. High local pH in pores, cracks and at grain boundaries and intergranular films has been found to promote selective dissolution and reprecipitation with enhanced local rates of attack. Consequently, hydrolysed products and TiO_2 precipitates, depending on the temperature of solution attack, are found preferentially at these points [15].

Hence, dissolution and surface reaction of this single-phase perovskite can be inhibited by (1) lowering pH to inhibit the base-catalysed hydrolysis reaction and stabilise the ion-exchanged layer; (2) maintaining a high concentration of the relevant dissolved cation close to the surface; (3) reduction of defects in the crystal lattice and as pores, microcracks, and intergranular films; and (4) increased grain size to reduce the proportion of the material as grain boundaries.

16.6.3 SILICON CARBIDE CERAMICS

Recent work on the corrosion behaviour of silicon carbide ceramics has suggested that similar mechanisms apply to this surface. The SiC was directly hydrolysed to a silica sol on its surface with its dissolution kinetics dependent on the sol stability [28]. Preferential attack was found at grain boundaries and around pores on the sample surface immersed in oxygenated alkaline solution. Acceleration of the mass loss with increasing pH is consistent with base-catalysed hydrolysis, and the $Si(OH)_4$ product equilibrates with partly dissolved species in solution [28].

The dissolution of single-phase and multiphase ceramics has recently been reviewed [11], including environmental degradation of ceramic superconductors based on the rare earth barium copper oxide perovskite-related layer structures with various levels of oxygen deficiency. Again, the mechanisms of surface degradation are similar to those found for the perovskite $CaTiO_3$, but the defect structure induced by the oxygen deficiency and the layer lattice of these compounds provide increased rates of surface degradation. Protection has been investigated using amorphous silica overlayers [29], silicon nitride formed by sol-gel [30] and nitrogen implantation [31] processes, deposition of metals [32], and the use of organic barrier layers [33–35]. It is possible that the same chemical control mechanisms as those used for perovskite may inhibit surface degradation, but research on this topic has not yet been undertaken.

16.6.4 SILICON NITRIDE OXIDATION

For the nitrides, like silicon nitride, passivation can be achieved by the normal process of oxide formation on the nitride surface. In fact, silicon nitride is often used as a passivation layer on silicon metal surfaces before the addition of other (e.g., metal) overlayers [21]. XPS spectra can be measured with different photoelectron exit angles, which, with constant electron mean free path in the solid, changes the effective escape depth so that, at small angles (e.g., 10°), almost all of the signal comes from the top monolayer and, at 90°, the maximum depth is sampled. Figure 16.4 is a set of Si 2p XPS spectra from silicon nitride with an oxide surface layer recorded as a function of photoelectron exit angle [21]. The photoelectron exit angle is measured relative to the plane of the sample surface. A curve fit of the 90° data illustrating two peaks corresponding to the nitride (i.e., 102 eV) and oxide (i.e., 103.8 eV) is indicated. It is clear that the ratio of oxide to nitride changes with depth, with the oxide signal predominating in the first monolayer and the nitride predominating in the 90° signal from a depth of five to seven atomic layers. Table 16.6 further illustrates this surface oxidation by comparing the relative intensities of the nitride and oxide peaks at different take-off angles. A parallel variation in the composition of the surface at different take-off angles is found for the O 1s and N 1s signals relative to the Si 2p signal. Hence, the oxide layer is relatively thin (i.e., <2 nm) and has the effect of passivating the surface and inhibiting further reaction.

Intergranular films, in addition to providing points of enhanced rates of attack, can determine the physical and mechanical properties of crystalline materials [36,37]. They provide regions into which specific elements pref-

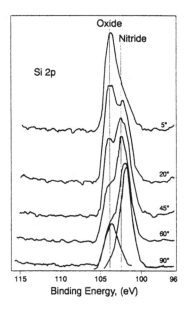

Figure 16.4 Silicon 2p spectra from surface-oxide–contaminated silicon nitride as a function of photoelectron take-off angle, measured relative to plane of the sample surface. The curve fit of the 90° data is included, while the line positions of the SiO_2 and Si_3N_4 compounds are indicated for the other take-off angles. (Reprinted with permission from Reference [21].)

erentially segregate and diffusion pathways for selective loss of particular elements. For instance, SEM and SAM examination of fracture faces of titanium boride ceramics have shown that carbon and boron distribute into the same areas in the intergranular regions, but the Ti, with Si and Ni impurities, is inhomogeneously dispersed [22]. These regions do not contain TiB_2, but distributions of the elements in chemical forms different from the bulk material. Hence, they provide regions of weakness and selective

TABLE 16.6. Silicon 2p: Curve Fit Intensities.

Take-Off Angle θ	Relative Intensities Si2p	
	Nitride	Oxide
5	28	2
20	49	51
45	64	36
60	70	30
90	76	24

reaction. In the second category, enhanced concentrations of alkali metal cations have been found at mineral grain boundaries, and these elements are lost to solution at rates exceeding those of other cations [12]. Static SIMS was used to obtain information on fracture face surfaces where the intergranular layers were found to be < 1 nm thick [38,39]. Imaging of intergranular films in alumina ceramics using high resolution TEM has shown similar elemental accumulation and selective reaction [40–42].

Other work in the Surface Technology Centre has found forms of surface degradation such as silicon carbide surface layers on vanadium carbide Toyota Diffusion–treated tools due to silica impurity, severe pitting due to pull-out of secondary phases in VC films, layering (with edges and corners exposed) of TiN films formed by PVD processes, and intermixing of silicon nitrides with metal surfaces, particularly titanium (see also References [43,44]). Oxidation at the interfaces between layers of metals, such as Al, on other substrates is also common, leading to detachment of the metal layer in certain circumstances. XPS with ion etching can detect the changes in composition of the metal and substrate with depths revealing these undesirable reaction products [22].

16.6.5 MULTIPHASE CERAMICS

For multiphase ceramics, the range of surface features discussed previously is further complicated by minor phases, precipitates and triple points, and preferential distribution of intergranular films between grains of different phase. Surface reactions of multiphase ceramics are similar to those in single-phase ceramics, but with particular complications from selective reaction. Relative phase durability and microencapsulation, together with selective reaction at the more complex microstructural features and selective precipitation and recrystallisation, complicate the description of surface degradation. For instance, in the multiphase ceramic Synroc C used for high-level nuclear waste disposal, it has been established that the following general order of reactivity applies to the phases: Mo alloy phases > perovskite > hibonite > hollandite $\gg$ zirconolite $\sim$ alumina $\gg$ TiO_2 [11,12]. Hence, surface exposed grains of the first two phases are preferentially lost to solution. The probability of continuing reaction then depends on whether these phases are microencapsulated or not, i.e., whether underlying grains of the same phase are connected at grain boundaries to the reacted grains or are isolated by intervening grains of more durable phases. In general, this depends on the volume fraction of the phase so that minor phases are microencapsulated, but a major phase, like perovskite, may continue to react. Limitation on the depth of reaction of the least durable major phase is then governed by the same considerations applying to single-phase ceramics, i.e., ion exchange, lattice hydrolysis, recrystallisation, precipitation, and the formation of ion-saturated protective layers. Hence, a very complicated, overlapped set of kinetics determines the observed rates of release of particular elements and surface degradation, particularly if an element is present in more than one phase. Examples of selective reaction of particular elements in Synroc C can be found in References [11,12,15]. In general, the combined information from single-phase studies with the concepts of selective reaction and microencapsulation allow interpretation of the observed surface degradation.

16.6.6 INTERGRANULAR FILMS

As with single-phase ceramics, the role of intergranular films can be critical. Static SIMS studies of these thin layers in titania- and alumina-based multiphase ceramics have suggested that these films predominate between grains of different phases and contain increased concentrations (over bulk phase values) of alkaline metal cations, silica, and alumina (apparently as aluminosilicates). Reaction via hydorlysis and diffusion is substantially enhanced in these layers. Surface analysis, using SAM, of the surfaces of closed pores exposed by fracture have also revealed one or two monolayers of Cs apparently condensed from gaseous cesium at the high temperatures used to form the ceramic [39]. These layers are removed immediately on contact with water, and it is possible that diffusion of this relatively free cesium via intergranular films, together with the cesium segregated into the films themselves, provides a significant proportion of continuing slow loss of this element to solution [11].

16.6.7 RECRYSTALLISATION AND PRECIPITATION

As with reaction of single-phase ceramics, hydrolysis products build up on the surface, and some dissolved species approach saturation in the solution or at the surface, resulting in precipitation and recrystallisation [11]. In some cases, hydrolysed products recrystallise *in situ* to form new phases (e.g., ZrO_2, TiO_2, Al_2O_3, iron oxyhydroxides, hydroxysilicates, and aluminosilicates) without dissolving [11]. Figure 16.5 illustrates the selective reaction and *in situ* recrystallisation of the perovskite phase in the multiphase ceramic Synroc C [12]. In other cases, precipitates such as $CaCO_3$, sulphates, molybdates, phosphates, etc., formed from saturated solution. The distribution of the former category is confined to the regions close to the phase from which the hydrolysis products were formed, but the latter precipitates tend to be distributed relatively uniformly across the surface. More detailed examples of these mechanisms, together with thermochemical modelling of dissolution and preciptiation behaviour based on solution analyses, can be found in the review of surface reactions and multiphase ceramics [11].

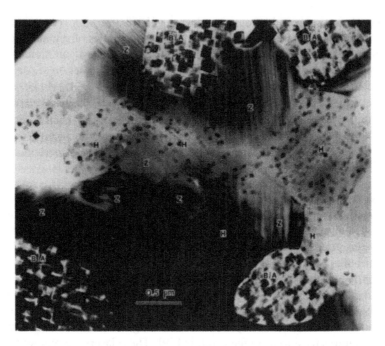

Figure 16.5 TEM micrograph of TiO₂ (brookite/anatase) crystals formed *in situ* from perovskite (CaTiO₃) grains in the surface of a Synroc B ceramic. An ion beam-thinned disc of Synroc B was subjected to hydrothermal attack in water at 190°C for one day. The original perovskite grain size was ca. 1 μm in diameter. The hollandite (H) and zirconolite (Z) phases are less reacted.

16.7 SURFACE DEGRADATION OF GLASSES

16.7.1 MECHANISMS

Mechanisms of corrosion of glasses have been extensively reviewed by Hench and Clark [19,20]. The range of glasses covers alkali borosilicate glasses through the more durable aluminosilicates to glass-ceramics used in waste disposal. This work has also emphasised the importance of combining surface and solution analysis in order to avoid misleading interpretations of results in inferring mechanisms (see also Reference [45]). In particular, inferences of congruent and incongruent dissolution based on solution analysis alone can be incorrect as shown in References [12,19]. Table 16.7 summarises the mechanisms of glass corrosion based on the extensive work in the literature.

16.7.2 EXAMPLES

Examples of the application of surface analytical techniques to the elucidation of these mechanisms can be found in a study of different glasses subjected to hydrothermal attack [12]. SIMS profiles from the surface of polished discs of glasses subjected to a high-temperature attack in distilled water have shown selective leaching of Cs, Na, and B to depths greater than 1 μm. Network attack and dissolution is found in infrared studies of the detached surface layer where systematic formation of SiOH groups with increased duration of attack and the loss of the boron-oxygen stretching vibration near 1400 cm^{-1} are features of the spectra. Solution analysis shows that several elemental species, e.g., Fe, Zn, Mg, Ti, after initial dissolution, show sharply decreasing loss rates with time. This is explained

TABLE 16.7. **Mechanisms of Glass Corrosion.**

1. **Ion exchange or selective leaching (incongruent dissolution)**

   Involves the exchange of mobile species, e.g., Na$^+$, K$^+$ Ca^{2+}, Mg^{2+} from the glass with protons or hydronium ions from the solution. Results in surface film formation.

2. **Network dissolution (congruent and surface film dissolution)**

   Involves breaking down of structural bonds in the glass or surface film ducts base catalysed hydrolysis. May occur uniformly or locally.

3. **Pitting**

   Localized network dissolution due to surface heterogeneities, stresses, or defects.

4. **Solution concentration**

   Involves the concentration of the solution with respect to species from the glass and reduction in concentration gradient. May result in a reduction of the corrosion rate.

5. **Precipitation**

   Involves the formation of insoluble compounds on the glass surface by reaction of the dissolved constituents from the glass with species already present in the solution. May be influenced by solution pH.

6. **Stable film formation**

   Involves the alteration of the glass surface composition by either diffusion processes or interfacially controlled reactions.

7. **Surface layer exfoliation**

   Involves the flaking off of surface layers formed by hydrolysis. Usually occurs after the glass has been removed from the solution due to dehydration and accompanying stresses.

8. **Weathering**

   Involves the interaction of humidity and reactive gases from the atmosphere with the glass surface. Usually results in the accumulation of precipitates (both soluble and insoluble) on the glass.

9. **Stress corrosion**

   Due to the mechanical abrasion and/or chemical reactions on the glass surface. Typical examples are high velocity wind or water carrying sand over glass surfaces.

by XPS and SIMS depth profiles in which these elements are found to accumulate in the first 200 nm of the reacted surface layer. In some cases
[46], crystalline phases form in the amorphous layer (e.g., zinc hemimor-
phite, wollastonite, and rhodonite), but in other glasses [12], accumulation
of these and other (e.g., Ni, Ce, and Nd) elements occurs in the reacted
surface layers with no crystalline forms evident in either XRD or electron
beam microdiffraction. Stable film formation and surface layer exfoliation
vary both with the composition of the glass and the nature of the surface
attack, particularly in wetting/drying weathering cycles in air [47]. Longer
term reaction of glass surfaces results in a variety of crystalline powdered
particles due either to precipitation of saturated species in solution and at
the exposed surface or to recrystallisation *in situ* [46,47]

16.7.3 CONTROL OF SURFACE DEGRADATION

Control of these surface reactions, therefore, lies in control of conditions of pH, temperature of solution, solution and surface concentrations
of exchangeable cations, diffusion barriers to movement of reactants (particularly OH^-) and products, and complexation of surface groups. These
principles have been used in recent studies of the inhibition of surface
reactions of granulated blast furnace slags [48].

16.8 SURFACE DEGRADATION OF POLYMERS

16.8.1 INTRODUCTION

The surface engineering of synthetic polymers by controlled degradation finds application in diverse areas, including interfacial adhesion of re-
inforcing fibres in composites [49,50], printing of hydrophobic surfaces
[51], bonding of polymers [51,52], and biocompatibility of polymers [53].
In addition, the degradation of the polymer surface is critical in the environmental performance of the material [54]. The list of engineering
polymers is growing, and the introduction of polymer alloys involving in-
terpenetrating networks, as well as phase-separated system [55], has meant
that the interfacial properties are often dominated by the surface energetics
of the system. The measurement of bulk properties such as mechanical
properties and molecular weight has been the traditional approach of the
polymer engineer, but more recently, it has been recognized that modern
surface analysis techniques provide the fundamental answer to the performance of the material in the above applications.

16.8.2 POLYMER STRUCTURE-PROPERTY RELATIONS

The majority of polymers used as engineering materials are based on a long-chain carbon backbone. Their physical properties vary widely but, to a first approximation, may be rationalized by noting the dependence on

(1) The length of the polymer chain (i.e., the molecular weight)
(2) The forces between the chains (controlled by the functional groups either in-chain or pendant)
(3) The nature and extent of crystallinity

An actual engineering material may consist of one or more polymers, as well as a range of fillers, and additives that are designed to modify the mechanical properties and confer stability to environmental degradation [56]. The processing of a polymer is also critical since the development of crystallinity depends on the thermal history of the material. Many polymers are semicrystalline due to the difficulty of achieving close packing of the long polymer chain, and the behaviour of the amorphous component is often the controlling factor in the performance. The thermal properties of a semicrystalline polymer may be described by the melting of the crystalline blocks (T_m) and by the glass transition of the amorphous region (T_g). These two temperatures generally define the properties since, below T_g, a polymer will be glassy and, between T_g and T_m, viscoelastic behaviour is observed. The energy absorbing properties of a polymer will depend on the ability of the polymer chain to deform, untangle, and dissipate energy, so if the polymer is used below T_g, it may fail in a brittle manner.

16.8.3 POLYMER DEGRADATION

The degradation of a polymer involves the loss of properties due to alteration of the physico-chemical structure by one or more of the following:

(1) Decreasing the length of the polymer chain by random chain scission, resulting in a loss of strength
(2) Changing the forces between chains by the incorporation of different functional groups such as oxidation products (This results in shrinkage and micro-crack formation.)
(3) Changing the extent of crystallinity due to recrystallization of polymer chains following scission
(4) Embrittlement due to an increase in T_g of the polymer following crosslinking between chains

In normal service, the environmental factors responsible for these changes in the polymer structure are UV (solar) radiation, heat, mositure, and chemical attack [54]. The nature and extent of degradation depends primarily on the chemical structure of the polymer. For example, polymers with photochemically labile groups (such as ketones) in the structure will be rapidly degraded by UV radiation; hydrolysis of polyesters by alkali and polyamides (nylons) by acids result from attack of functional groups in the polymer leading to chain scission.

In most cases, the degradation of the polymer initiates at the surface. This is principally due to the high concentration of the reactant (such as oxygen for an oxidation process) but also may result from the presence of defect groups or morphological differences due to processing. The stress state of the polymer can also affect the rate of chemical reactions. This may be seen from the environmental performance of nylon 6 when exposed outdoors under mechanical load [57]. As shown in Figure 16.6, there is a rapid decrease in the fracture energy, which is more rapid at the higher stress. Scanning electron micrographs (Figure 16.7) show that, initially, a network of fine cracks is developed, then followed by surface erosion. The cracks occur perpendicular to the stress direction. Figure 16.8 shows micrographs from a sample exposed outdoors without stress, and it can be seen that microcracks develop initially in one direction [Figure 16.8(a)], and thereafter larger cracks grow in a random manner with no evidence for

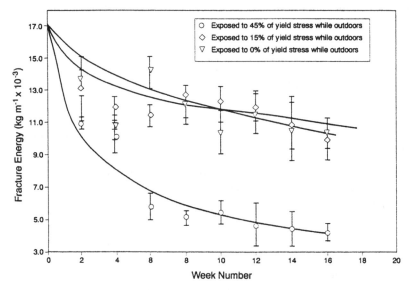

Figure 16.6 The change in fracture energy as measured by the area under the stress-strain curve for nylon 6.

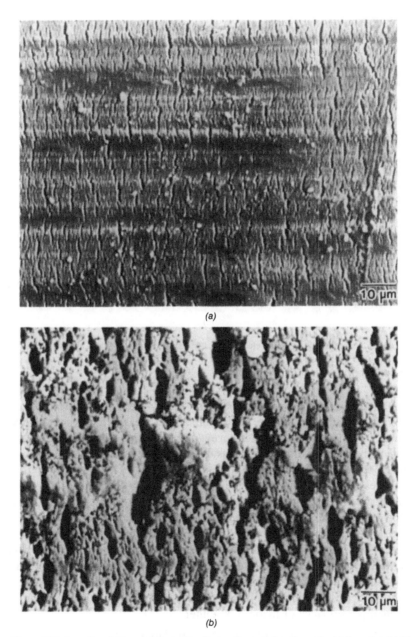

(a)

(b)

Figure 16.7 Scanning electron micrographs of nylon 6 exposed outdoors for (a) 5 weeks, (b) 10 weeks, and (c) 20 weeks at a load of 45% of yield stress. Crack formation has occurred perpendicular to the stress direction.

297

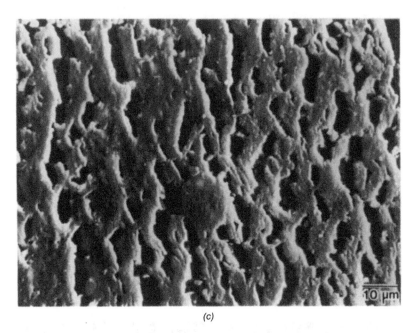

(c)

Figure 16.7 (continued) Scanning electron micrographs of nylon 6 exposed outdoors for (a) 5 weeks, (b) 10 weeks, and (c) 20 weeks at a load of 45% of yield stress. Crack formation has occurred perpendicular to the stress direction.

the surface erosion previously noted in highly stressed samples (Figure 16.7) in the same time frame. Molecular weight measurements showed that chain scission occurred on weathering and this was accelerated by the applied stress. X-ray diffraction measurements showed that recrystallisation of these degraded polymer chains was occurring. The resultant density changes and restructuring of the previously amorphous fraction of the semicrystalline polymer was the probable precursor to microcrack formation at the surface, which resulted in a lowering of the energy to fracture the polymer. The polymer chain scission resulted from an oxidative chain reaction initiated by UV radiation (as described in the next section). The surface erosion was not reported when nylon 6 was exposed in a hot, dry climate, and the rate of loss of fracture stress was accordingly lower [58]. This indicates that moisture has a significant role to play in the environmental performance of nylon 6 by either

(1) Removing low molecular weight degradation products
(2) Disrupting the hydrogen bonding and affecting mechanical properties
(3) Swelling the polymer, so enabling surface recrystallisation

In all cases, it is the chemistry at the surface that controls the ultimate performance of the polymer.

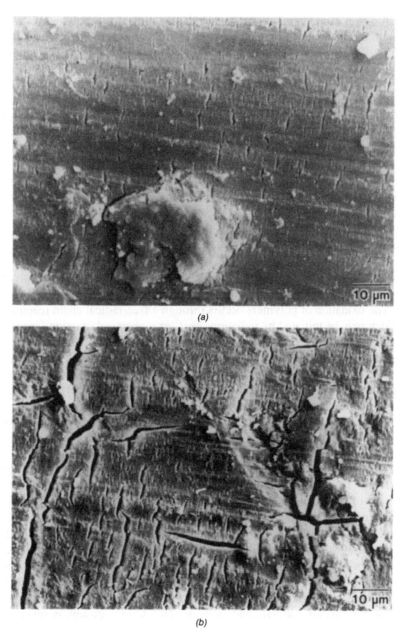

(a)

(b)

Figure 16.8 Scanning electron micrographs of nylon 6 exposed outdoors without mechanical load for (a) 5 weeks and (b) 10 weeks.

A closer examination of the surface degradation of this polymer showed that the processing cycle had an important role to play in the performance [59]. When nylon 6 is injection moulded, the temperature of the mould will affect the rate of cooling and thus surface morphology. Quenching resulted in a thicker nonspherulitic zone at the surface, and this showed a lower extent of surface cracking and erosion during outdoor exposure under load than from an annealed surface. The process cycle also affected the creep of the viscoelastic polymer, and it is possible that the higher creep of the quenched material resulted in chain orientation, which lowered the rate of oxidation of the surface [59]. This change in the surface morphology thoughout the process cycle of the polymer explains, in part, the large scatter in properties observed from injection moulded polymers when exposed outdoors.

16.8.4 POLYMER OXIDATION

The degradation processes described previously have generally involved a change in the polymer structure due to oxidation.

The oxidation of polymers occurs through a free-radical chain reaction, which may be initiated by radiation, heat, or a chemical reaction (such as decomposition of a peroxide or other labile molecule). In a solid, semi-crystalline polymer (such as polyethylene), the depth to which oxidation occurs is often controlled by the solubility and diffusion coefficient of oxygen in the polymer. This, in turn, may be affected by the crystallinity and extent of orientation of the polymer.

While extensive oxidation of a polymer results in scission of the polymer chain, leading to microcracking, embrittlement, and a loss of strength as described in the previous section, controlled surface oxidation is frequently used to enhace the wettability and adhesive bonding of the surface. For example, the surface energy of polypropylene may be increased from 30 to 38 mN m^{-1} by a simple flame treatment. This is sufficient to render the polymer printable. This oxidation leads to the introduction of a range of functional groups such as ether, alcohol, carbonyl, acid, ester, and hydroperoxide. Some of these represent the end-product of the oxidation, while others (notably the hydroperoxide) are intermediates that may be unstable and lead to chain branching in the oxidation sequence. A simplified oxidation sequence that shows the formation of hydroperoxide (ROOH), alcohol (ROH), and carbonyl ($R_2'' C=O$) groups is given in Table 16.8.

To obtain more detailed information of the chemistry at the surface of a polymer during degradation, surface-specific spectroscopy must be employed. FT-IR has been widely used for the "near-surface" analysis of the first 0.3–10 μm of the polymer by using Attenuated Total Reflectance

TABLE 16.8. Simplified Free Radical Reaction Scheme for the Oxidation of a Hydrocarbon Polymer.

Initiation	ROOR′	$\xrightarrow[\text{heat}]{\text{catalyst}}$	RO˙R′O˙ ⎫		
	RO˙ + RH	$\longrightarrow$	R˙ + ROH ⎭	r_i	(1)
Propagation	R˙ + O₂	$\xrightarrow{k_2}$	RO₂˙		(2)
	RO₂˙ + RH	$\xrightarrow{k_3}$	ROOH + R˙		(3)
Termination	R˙ + R˙	$\xrightarrow{k_4}$	R—R		(4)
	RO₂˙ + RO₂˙	$\xrightarrow{k_5}$	ROH + O₂ + R₂″C=O		(5)

(ATR) methods. In this technique, the sample is placed against a high refractive index optical element such as KRS-5 or germanium, and the analysing infrared beam is passed through it as shown in Figure 16.9. Multiple internal reflections occur at the interface between the element and the polymer, and the resultant evanescent wave setup at the interface penetrates into the polymer for a distance d, which depends on the respective refractive indices of the polymer (n_1) and optical element (n_2); the angle of incidence, θ; and the wavelength, λ, of the radiation [60]:

$$d = \frac{\lambda}{2\pi n_1 (\sin^2 \theta - n_1)^{1/2}}$$

The resultant ATR spectrum, therefore, differs from a transmission spectrum as the sampling depth varies with wavelength and is biased to longer wavelengths. Another limitation is that good contact must be achieved between the sample and ATR element, which, in degraded, cracked surfaces such as those shown in the electron micrographs of nylon 6 (Figures 16.7 and 16.8), is clearly difficult.

In spite of these limitations ATR–FT-IR has provided valuable information about the chemical species in the near-surface region of oxidized polyolefins such as polypropylene [61]. By using optical elements of different refractive index and changing the angle of incidence, θ, an estimate of the product distribution over an order of magnitude difference in depth may be obtained. The sensitivity and specificity of FT-IR may be enhanced by selective derivatization of the oxidation products. It has been demonstrated, by using nitric oxide as a reagent, that it is possible to dis-

IR in   Polymer

IR out

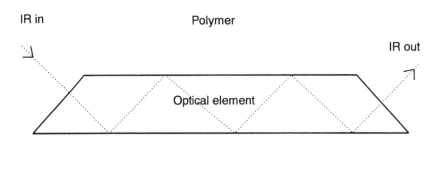

Polymer

Figure 16.9 Measurement of infrared spectrum of a polymer surface by Attenuated Total Reflectance using a Germanium or KRS-5 Optical Element.

tinguish different types of hydroperoxides along the polymer backbone – critical information in understanding the initiation of degradation at the polymer surface [62].

While FT-IR has been valuable in relating the chemistry of the near-surface to structural changes in the material during environmental degradation, for many studies such as adhesion and biocompatibility of polymers, actual surface information in the first 1–10 nm is required. The analysis of functional groups resulting from oxidation reactions in this top layer is the goal of much polymer surface analysis. The modern surface analysis techniques already described for the study of metals, ceramics, and glasses are readily adaptable to the study of the surface structure of polymers. XPS has been the most widely emplyed [63], but it has been shown that some modification of the technique, notably surface derivatization [51,52,60,64], may be required in order to analyse the wide range of groups described previously. However, there are particular requirements that must be met because of the inherent instability of some polymers to radiation and ion sources, as well as the surface restructuring that occurs in a viscoelastic material [60]. These particular requirements for successful polymer surface analysis will be discussed in the context of examples that show the power of X-ray photoelectron spectroscopy and static secondary ion mass spectrometry in solving specific surface engineering problems.

16.8.5 XPS OF OXIDIZED POLYMERS

One example of controlled surface engineering by selective oxidation is the enhancement of chemical bonding of ultra high modulus polyethylene fibres (UHMPE) to a crosslinked resin to form a fibre-reinforced composite [49,50]. This may be achieved by (1) a corona discharge treatment in air, (2) a microwave or shortwave plasma treatment in oxygen, or (3) a chemical oxidation reaction.

A measure of the effectiveness of the treatment is the increase in interfacial strength as measured by fibre pull-out force. However, this gives no information as to the surface chemistry leading to this improvement. XPS of the surfaces treated above show differences in the surface oxidation products. A typical survey and multiplex spectrum of oxidized polyethylene [50] is shown in Figure 16.10. These show the typical features of the C1s region of an organic polymer:

(1) Surface charging of the highly insulating polymer shifts the binding energy by ~2 eV to higher energy of the C1s band at 284.6 eV.

(2) The tail to higher binding energy contains the spectral information of the oxidation products with the bands shifting ~1.5 eV for each single $C-O$ bond [63]. Thus, based on C1s of 284.6 eV, it is expected to be

$$C-O: 286.1 \text{ eV}$$
$$O-C-O \text{ or } C=O: 287.6 \text{ eV}$$
$$O-C=O: 289.2 \text{ eV}$$

(3) The overlapping bands in the C1s tail require curve fitting in order to deconvolute the spectrum and assign the relative amount of each functional group present. Successful curve fitting of the broad bands obtained with a nonmonochromatic X-ray source (e.g., $MgK\alpha$) may only be achieved by adopting a strict protocol with a close tolerance of peak position and band width. A band width of 1.7 eV is appropriate in order to allow for peak broadening effects during oxidation.

The multiplex spectrum in Figure 16.10 has been curve fitted using this protocol, and it is seen that it may be resolved into four peaks consisting of each of the functional groups shown previously. By using the areas of each peak in an XPS spectrum and the sensitivity factors from standard compounds, the surface atomic concentrations may be obtained. A plot for the oxidized species on the surface of polyethylene fibres subjected to different oxidation treatment is shown in Figure 16.11 and allows the extent of secondary oxidation to be assessed. This method does not allow several similar functional groups to be distinguished (e.g., $O-C$ and $C-O-O$) and methods based on surface derivatization have been proposed [51,52]. Some suitable reagents, their products, and the XPS line analysed are shown in Table 16.9. It is noted that the analysis is now shifted from carbon to another element, so the overlapping bands are no longer a problem.

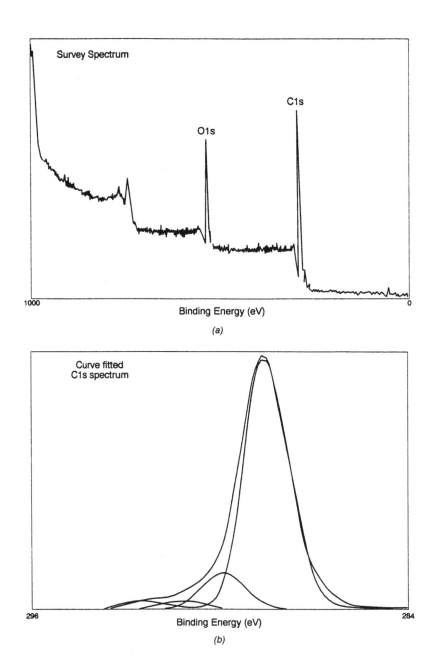

Figure 16.10 XPS survey spectrum and curve fitted multiplex spectrum for plasma-treated Ultra High Modulus Polyethylene (UHMPE).

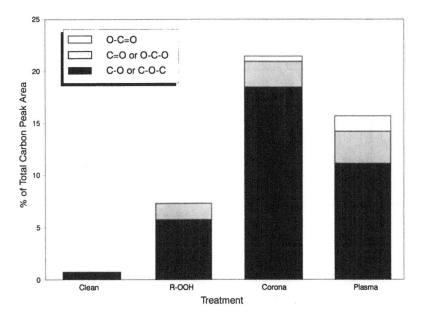

Figure 16.11 Product distribution on the surface of ultra high modulus polyethylene fibres subjected to different surface oxidation treatments (R−OOH = surface hydroperoxidation).

Similar derivatization reactions have been used in FT-IR analysis of oxidized polymers [62], and it has been shown that it is extremely difficult to obtain reproducible derivatization with other than gas phase reactants. In addition, the use of any reagent for quantitative surface analysis assumes complete reaction and no surface restructuring when the reagent diffuses into the top 50–100 Å of the oxidized polymer. In particular, the latter problem may occur with liquid phase reagents [60]. It has also been noted that certain polymer additives may react with the reagent, so falsifying the analysis [60].

16.8.6 X-RAY RADIATION DAMAGE DURING XPS ANALYSIS

It has generally been assumed that radiation damage to polymers during XPS is negligible since the main elemental line intensities do not vary on repeated analyses. However, in the case of oxidized polymers, which are subjected to prolonged multiplex analysis in order to measure depth profiles by comparing grazing and normal angle XPS signals (as discussed below), considerable damage to certain functional groups may occur.

An example is the analysis of the plasma and corona discharge oxidized polyethylene fibre, as shown in Figures 16.10 and 16.11. The total O1s signal

TABLE 16.9. Some Derivatization Reactions for Studying Oxidized Polymer Surfaces and the XPS Line Analysed [3,16].

Group	Reagent	Product	XPS Line
—C— ‖ O (Carbonyl)	$C_6H_5NHNH_2$	$\overset{\|}{C}{=}N{-}NC_6H_5$	F 1s
—C— | O—OH (Hydroperoxide)	SO_2	$\overset{\|}{\underset{\|}{C}}{-}OSO_2OH$	S 2p
—C=O | OH (Carboxyl)	NaOH	$\overset{\|}{\underset{O}{C}}{-}O\bar{N}a^+$	Na 1s

from the plasma-treated polymer was found to decrease with increased analysis time. This was traced to the radiation sensitivity of the carboxyl group, a prominent plasma oxidation product. The carboxyl group under X-ray irradiation decomposes to carbon dioxide, so that short analysis times were required for functional group quantitation with the consequent loss in signal-to-noise. The dose rates in XPS (which may reach over 10 Gy min^{-1} with nonmonochromatic sources) are clearly sufficient to affect the analysis of oxidized polymers. Similar problems have also been reported in the analysis of halogenated polymers due to dehalogenation in the X-ray beam [60]. This problem does not generally occur in the analysis of metals, ceramics, and glasses, and it is important to recognize the inherent instability of many polymers to high energy radiation.

16.8.7 DEPTH PROFILING OXIDIZED POLYMERS

In many applications involving polymer oxidation, it is important to know the depth to which oxidation has occurred. In the surface analysis of metals and ceramics, this is most readily achieved by ion etching the surface at a known rate with periodic XPS analysis. By using secondary ion mass spectrometry (SIMS), surface structural information may also be obtained from analysis of the positive and negative ions sputtered by the primary ion beam. At the dose rates necessary for etching, the polymer surface is immediately degraded, so a different strategy is required.

One of the simplest methods of depth profiling flat surfaces is to measure the XPS signal at different emergent angles. Because of the limited escape depth of the photoelectron ($\sim 3\lambda$ is the inelastic mean free path), spectra obtained at an increased angle, θ, with respect to the normal, will be from closer to the surface. This may be used to detect an overlayer by examining the change in the XPS signal of an element that is not oxidized (e.g., nitrogen in a nylon polymer) as the take-off angle is varied [63,65].

This is less successful with curved surfaces such as fibres, and an alternative approach uses the kinetic energy dependence of the electron escape depth. The mean free path for organic compounds has been found to fit an empirical equation of the form:

$$\lambda = 49\ E^{-2} + 0.11\ E^{1/2}$$

where E is the kinetic energy of the photoelectron [66]. Some idea of the depth of an oxidized layer can be obtained either by

(1) Comparing the intensity of signals from photoelectrons having different kinetic energies but from the same element (e.g., O1s and O2s)
(2) Using X-ray sources of higher energy [e.g., Ti ($K\alpha$) = 4510 eV] (This anode would give a sampling depth of ~ 300 Å for C1s compared to 40 Å for the usual Mg ($K\alpha$) anode [60].)

16.8.8 SIMS OF DEGRADED POLYMERS

In contrast to the study of metals, nonmetals, and glasses, secondary ion mass spectrometry (SIMS) of polymers is most successfully carried out under static, rather than dynamic, conditions. In order to achieve high analytical sensitivity, high beam currents are employed in the latter materials, resulting in several monolayers per secend being removed from the surface. This dynamic SIMS has the additional advantage of producing a depth profile for the elemental composition as the surface is etched away. These conditions, characterised by primary beam currents of several μA cm^{-2}, result in immediate degradation of the polymer surface so that etch rates are uncertain and the SIMS spectrum does not reflect the surface composition. However, if the primary beam current is reduced to around 1 nA cm^{-2}, the etch rate is less than one monolayer per hour. Under these static SIMS conditions, the spectrum is truly representative of the surface and there is minimal beam damage. This is ideal for polymer surface analysis, and there has been much recent research on the application of the technique to polymers [67]. Several problems have been recognized that may limit the quantitative analysis:

(1) Time dependence of peak intensities of certain mass fragments due to ion beam damage of aromatic groups

(2) Changes in the surface potential due to charge build-up (This moves the position of peaks with respect to the energy acceptance window of the quadrupole mass spectrometer, resulting in a change in the relative intensity of the ions.)

It is considered that ion beam damage is the more important problem, and the lowest beam current for the shortest time should be used. This is important for SIMS imaging, which has been applied to polymers using source spot sizes of 1 μm or less [67]. Under these conditions, for high magnification, sufficient ion yield for analysis requires a high local dose. This means that several monolayers may be lost and static SIMS conditions no longer apply.

An early example, which showed the high sensitivity and structural information that could be obtained by static SIMS of a degraded polymer surface, was the study of the hydrolysis of poly-t-butyl methacrylate under mild conditions [68]. By following the disappearance of the mass spectral fragment corresponding to the t-butyl group, it was possible to observe ester hydrolysis where no change could be detected by XPS.

Static SIMS and XPS are complementary techniques for studying polymer surfaces. XPS provides quantitative elemental composition and bonding information, while SIMS provides analysis of the monolayer at the surface but only limited absolute quantitative information.

16.9 REFERENCES

1 Hirschwald, W. "Selected Experimental Methods in the Characterisation of Oxide Surfaces," in *Surface and Near-Surface Chemistry of Oxide Materials,* J. Nowotny and L. C. Dufour, eds., Elsevier, 1988.

2 Briggs, D. and Seah, M. P., eds. *Practical Surface Analysis.* John Wiley, New York, 1987.

3 Woodruff, D. P. and Delchar, T. A. *Modern Techniques of Surface Science.* Cambridge Univ. Press, 1986.

4 Benninghoven, A., Rudenower, F. G. and Werner, H. W. *Secondary Ion Mass Spectrometry, 86, Chemical Analysis Series,* P. J. Elwing, J. D. Winefardner and I. M. Kolthoff, eds., John Wiley, New York, 1987.

5 Hofmann, S. *Surf. Interface Anal.,* 9:3, 1986.

6 Brown, A. and Vickerman, J. C. *Surf. Interface Anal.,* 6:1, 1986.

7 Willis, R. F., ed. *Vibrational Spectroscopy of Adsorbates.* Springer, Berlin, 1980.

8 Ferraro, J. R. and Krishnan K., eds. *Practical Fourier Transform Infrared Spectroscopy.* Academic Press, San Diego, California, 1990.

9 Wolery, T. J. "Calculation of Chemical Equilibria between Aqueous Solution and Minerals," The EQ3/6 Softward Package, Lawrence Livermore Lab., Rep. UCRL-52658, 1979.

10 Segall, R. L., Smart, R. St. C. and Turner, P. S. "Oxide Surfaces in Solution," Chapter 13 in *Surface and Near Surface Chemistry of Oxide Materials,* L. C. Dufour and J. Nowotny, eds., pp. 527–576, Elsevier, 1988.

11 Myhra, S., Pham, D. K., Smart, R. St. C. and Turner, P. S. "Surface Reactions and Dissolution of Ceramics and High Temperature Superconductors," in *Surface and Near-Surface Chemistry of Ceramics,* J. Nowotny, ed., Elsevier Science, Amsterdam, 1991.

12 Myhra, S., Smart, R. St. C. and Turner, P. S. "The Surfaces of Titanate Minerals, Ceramics and Silicate Glasses: Surface Analytical and Electron Microscope Studies (Review)," *Scanning Microscopy,* 2:715–734, 1988.

13 Jones, C. F., Segall, R. L., Smart, R. St. C. and Turner, P. S. "Initial Dissolution Kinetics of Ionic Oxides," *Proc. Roy. Soc. A,* A374:141–153, 1981.

14 Jones, C. F., Segall, R. L., Smart, R. St. C. and Turner, P. S. "Surface Structure and the Dissolution Rates of Ionic Oxides," *J. Materials Sci. Letters,* 3:810–812, 1984.

15 Turner, P. S., Jones, C. F., Myhra, S., Neall, F. B., Pham, D. K. and Smart, R. St. C. "Dissolution Mechanisms of Oxides and Titanate Ceramics – Electron Microscope and Surface Analytical Studies," in *Surfaces and Interfaces of Ceramic Materials,* L. C. Dufour et al., eds. Kluwer Academic Publishers, pp. 663–690, 1989.

16 Segall, R. L., Smart, R. St. C. and Turner, P. S. "Ionic Oxides – Distinction between Mechanisms and Surface Roughening Effects in the Dissolution of Magnesium Oxide," *J. Chem. Soc., Faraday I,* 74:2907–2912, 1978.

17 Pease, W. R., Segall, R. L., Smart, R. St. C. and Turner, P. S. "Comparative Dissolution Rates of Defective Nickel Oxides from Different Sources," *J. Chem. Soc. Faraday I,* 82:759–766, 1986.

18 Petit, J. C., Dell Mea, G., Dran, J. C., Magonthier, M. C., Mando, P. A. and Paccagnella, A. "Hydrated-Layer Formation during Dissolution of Complex Silicate Glasses and Minerals," *Geochimica et Cosmochimica Acta,* 54:1941– 1955.

19 Hench, L. L., Clark, D. E. and Yen-Bower, E. Lue. "Surface Leaching of Glass and Glass-Ceramics," *Nuc. Chem. Waste Man.,* 59:1–39, 1980.

20 Clark, D. E. and Hench, L. L. "An Overview of the Physical Characterisation of Leached Surfaces," *Nucl. Chem. Waste Man.,* 2:93–101, 1981.

21 Bomben, K. D. and Stickle, W. F. *Surface Analysis Characterisation of Thin Films, Microelectronic Manufacturing and Testing.* Lake Publishing, Libertyville, Illinois, 1987.

22 Smart, R. St. C. "Surface Characterisation of Electrodes," in *New Developments in Electrode Materials and Their Applications,* A. J. Jones and L. Wood, eds., Dept. Industry, Technology and Commerce, Aust. Govt. Press, pp. 101–115, 1990.

23 Pease, W. R., Segall, R. L., Smart, R. St. C. and Turner, P. S. "Evidence for Modification of Nickel Oxide by Silica: Infrared, Electron Microscopy and Dissolution Rate Studies," *J. Chem. Soc. Faraday I,* 76:1510–1519, 1980.

24 Smart, R. St. C., Baker, B. G. and Turner, P. S. "Aspects of Surface Science in Australian Industry," *Aust. J. Chem.,* 43:241–256, 1990.

25 Pham, D. K., Neall, F. B., Myhra, S., Smart, R. St. C. and Turner, P. S. "Dissolution Mechanisms of $CaTiO_3$ – Solution Analysis, Surface Analysis and Electron Microscope Studies – Implications for Synroc," *Mat. Res. Soc. Symp. Proc.,* 127:231–240, 1989.

26 Myhra, S., Pham, D. K., Smart, R. St. C. and Turner, P. S. "Dissolution

Mechanisms of the Perovskite and Hollandite Phases in the Synroc Assemblage, Scientific Basis of Nuclear Waste Management XIII," *Mat. Res. Soc. Symp. Proc.*, 1990.

27 Kastrissios, T., Stephenson, M., Turner, P. S. and White, T. J. "Hydrothermal Dissolution of Perovskite: Implications for Synroc Formulation," *J. Amer. Ceram. Soc.*, 70:C144–146, 1987.

28 Hirayama, H., Kawakubo, T. and Goto, A. "Corrosion Behavior of Silicon Carbide in 290°C Water," *J. Am. Ceram. Soc.*, 72(11):2049–2053, 1989.

29 Baney, R. H., Bergstrom, D. F., Carpenter, L. E., Petersen, D. R., Elwell, D. F., Shapiro, A. A. and Fleishner, P. S. *Mater. Res. Symp. Proc.*, 176, 1990.

30 Jia, Q. X. and Anderson, W. A. *J. Appl. Phys.*, 66:452, 1989.

31 Chaudhari, S. M., Viswanathan, R., Bendre, S. T., Nawale, P. P. and Kanetkar, S. M. *J. Appl. Phys*, 66:4509, 1989.

32 Chang, C. *Appl. Phys. Lett.*, 53:1113, 1988.

33 Jin, S., Liu, L., Zhu, Z. and Huang, Y. *Solid State Communications*, 69:179, 1989.

34 Sato, K., Omae, S., Kojima, K., Hashimoto, T. and Koimuma, H. *Japan J. Appl. Phys.*, 27:L2088, 1988.

35 Barns, R. L. and Laudise, R. A. *Appl. Phys. Lett.*, 51:1373, 1987.

36 Chadwick, G. A. and Smith, D. A., eds. *Grain Boundary Structure and Properties*. Academic Press, 1976.

37 Ralph, B. and Porter, A. "The Study of Grain Boundary Structure and Properties in Engineering Materials," *J. Microsc. Spectrosc. Electron.*, 8:159–175, 1983.

38 Smart, R. St. C. "The Validity of SIMS Observations of Alkali Metal Segregation into Intergranular Regions of Ceramics," *Appl. Surface Sci.*, 22/23:90–99, 1985.

39 Cooper, J., Cousens, D. R., Hanna, J. A., Lewis, R. A., Myhra S., Segall, R. L., Smart, R. St. C., Turner, P. S. and White, T. J. "Intergranular Films and Pore Surfaces in Synroc C: Sturcture, Composition and Dissolution Characteristics," *J. Amer. Ceram. Soc.*, 69:347–352, 1986.

40 Jantzen, C. N., Clarke, D. R., Morgan, P. E. D. and Harker, A. D. *J. Amer. Ceram. Soc.* 65:292, 1982.

41 Clarke, D. R. *J. Amer. Ceram. Soc.*, 64:89, 1981.

42 Clarke, D. R. *Ultramicroscopy*, 8:95, 1982.

43 Shichi, Y., Matsukiyo, K. and Matsunaga, M. "SIMS Analysis of Joining Interface between Silicon Nitride and Metal," *Bunseki Kagaku*, 38:505–509, 1989.

44 Shichi, Y., Inoue, Y. and Matsunaga, M. "SIMS Analysis of the Joining Interface between Silicon Nitride and Metal Using a Caesium Ion Source," *Bunseki Kagaku*, 39:T93–T98, 1990.

45 Cousens, D. R., Lewis, R. A., Myhra, S., Segall, R. L., Smart, R. St. C. and Turner, P. S. "Evaluating Glasses for High Level Radioactive Waste Immobilization—A Review," *Radioactive Waste Management*, 2(2):143–168, 1968.

46 Lewis, R. A., Myhra, S., Segall, R. L., Smart, R. St. C. and Turner, P. S. "The Surface Layer Formed on Zinc-Containing Glass during Aqueous Attack," *J. Non Cryst. Solids*, 53:299–313, 1982.

47 Clark, D. E. and Yen-Bower, E. L. "Corrosion of Glass Surfaces," *Surf. Science*, 100:53–70, 1980.

48 Abo-El-Enien, S. A., Daimon, M., Ohsawa, S. and Kondo, R. "Hydration of Low Porosity Slag-Lime Pastes," *Cement and Concrete Research*, Vol. 4, No. 2. Pergamon Press Inc., 1974.

49 Nardin, M. and Ward, I. M. *Mat. Sci. Technol.*, 3:814, 1987.

50 George, G. A. and Willis, H. A. *High Perf. Polym.*, 1:335, 1989.

51 Briggs, D. *Polymer,* 25:1379, 1984.

52 Gerenser, L. J., Elman, J. F., Mason, M. G. and Pochan, J. M. *Polymer,* 26:1162, 1985.

53 Andrade, J. *Surface and Interfacial Aspects of Biomedical Polymers.* Plenum Press, New York, 1985.

54 Davis, A. and Sims, D. *Weathering of Polymers.* Applied Science, UK, 1983.

55 MacKnight, W. J. and Karasz, F. E. "Polymer Blends," in *Comprehensive Polymer Science,* G. Allen, ed., Pergamon, UK, 7:111, 1989.

56 Scott, G. *Pure Appl. Chem.,* 30:267, 1972.

57 George, G. A. and O'Shea, M. S. *Mater. Forum,* 13:11, 1989.

58 Qayyum, M. M. and White, J. R. *J. Mater. Sci.,* 21:2391, 1986.

59 George, G. A. and O'Shea, M. S. *Polym. Deg. Stabil.,* 28:289, 1990.

60 Gillberg, G. *J. Adhesion,* 21:129, 1987.

61 Carlsson, D. J. and Wiles, D. M. *J. Macromol. Sci.-Rev. Macromol. Chem.,* C14:65, 1976.

62 Carlsson, D. J., Brusseau, R., Zhang, C. and Wiles, D. M. *Polym. Deg. Stab.,* 17:308, 1987.

63 Dilks, A. Chapter 4 in *Dev. Poly. Characterization −2,* J. V. Dawkins, ed., Applied Science, UK, 1980.

64 Babitch, C. D. *Appl. Surf. Sci.,* 32:57, 1988.

65 Chappell, P. J. C., Williams, D. R. and George, G. A. *J. Coll. Inter. Sci.,* 134:385, 1990.

66 Seah, M. P. and Dench, W. A. *Surf. Inter. Anal.,* 1:2, 1979.

67 Vickerman, J. C. "Static Secondary Ion Mass Spectrometry," in *Methods of Surface Analysis−Techniques and Applications,* J. M. Walls, ed., Cambridge UP, UK, 1989.

68 Gardella, J. A., Novak, F. P. and Hercules, D. M. *Anal. Chem.,* 56:1371, 1984.

CHAPTER 17

The Effectiveness of Non-Destructive Techniques for Assessment of Quality of Plasma-Sprayed Coatings

M. E. HOUGHTON[1]

17.1 INTRODUCTION

THIS chapter gives an overview of some of the more promising contemporary research into nondestructive testing methods, which may have potential for the assessment of the "quality" (or "integrity") and thickness of plasma-sprayed coatings.

Through the incorporation of sophisticated electronic and microprocessor control technologies, plasma-spray processes are becoming more capable of integration into advanced, automated, and controlled manufacturing systems. An example of this greater degree of sophistication is the recent development of an Electronic Plasma Gun by the CSIRO Division of Manufacturing Technology [1]. This spray torch has enhanced electronic control of the plasma and, by using a unique current pulsing technique, provides a much better control over spraying conditions. Besides these technological developments, there is a need for a more general acceptance of quality assurance techniques by the spraying industry, and this trend is leading to tangible benefits for the sprayed component consumer.

A comprehensive review of the automation and control of plasma-spray processes has recently been made [2], and from this study, it is clear that, despite the undoubted contemporary technological and process control advances that have taken place, there remains the major problem of how to assess a coated product in an in-line day-by-day production situation. Several methods are currently available for inspecting and assessing the quality of coated product samples [3]. However, the majority of

[1]CSIRO, Division of Manufacturing Technology, Melbourne Laboratory, Locked Bag No. 9, Preston, Victoria 3072, Australia.

these—particularly those that have become international standards [4–8]—are destructive in nature, and consequently not applicable for use in a production-line or in-service 100% inspection situations. Most methods involve examinations of "representative" specimens using microscopic (metallographic), mechanical property (tensile, bending, fracture mechanics, scratch, or indentation hardness), crystallographic (X-ray diffraction), and/or coulometric (chemical dissolution) assessment techniques. Each method usually requires access to the facilities or resources of a suitably equipped quality or research laboratory. By contrast, researchers in NDT are seeking an appropriate method(s) of coating assessment, which is reliably accurate, rapid, noncontact, and can be applied to components that may be both remote from, or within, the extremely noisy and electronically inclement spray booth environment. An additional objective, particularly from the CSIRO viewpoint, is that an ideal NDT technique should be one that is capable of being incorporated along with the EPG into an integrated, automated, and controlled manufacturing system. It would seem that this ideal is still a considerable way from commercial reality.

17.2 THE "QUALITY" OF A SPRAYED COATING

The "quality" of a sprayed product is invariably a compromise. Ever-increasing demands for more severe service conditions, as well as the control of in-process conditions—technical, environmental, safety, and economic—can all influence quality. The attainment of satisfactory quality for plasma-sprayed coatings will be a function of good process control at every stage of the processing cycle, and it is only through strict adherence to tight process control procedures that "reliable" deposits of "acceptable" quality can be produced.

Various alternative combinations of properties, parameters, or characteristics may be used to define and/or determine the quality of a coating. Some of these features are

- thickness
- covering power
- surface finish
- porosity or density
- internal stress distribution
- adhesion to substrate
- chemical reactivity
- electrical properties
- wear resistance

- evenness
- edge definition
- microstructure
- hardness
- cohesive strength
- chemical composition
- thermal properties
- frictional properties
- machineability

The actual combination of these features, which most accurately describe the quality (or integrity) of a coating for a given application, can at times be critical, as the attainment of one property level might be at the expense of another. The precise criteria to be used will tend to depend on the application, and it is important that "customer and producer agree on the properties by which the quality will be judged" [9].

Perhaps the most common defects present in a spray coating, which require rapid identification and eradication, include

- inefficient adhesion/bonding at the coating/substrate interface
- the presence of cracks in the coating parallel to the substrate (including the potential for delamination)
- an unusually high inclusion content
- the presence of cracking normal to the substrate
- the presence of unmelted particles
- "glazing" of presolidified strata within, or on the surface, of the coating, due to interruption in powder feeding
- an unusually high presence (or distribution) of voids or porosity
- excessive residual stress

These various flaws and microstructural characteristics may be "manifest as a desirable combination of service functions" [10]. In such a case, the coating may be said to possess satisfactory integrity. Under these circumstances, the coating's integrity will include its ability to stick to the substrate by whatever means (this may also be termed its *bond integrity* or *adhesion strength*), as well as embrace other desirable features such as low porosity and uniformity of composition.

The purpose of any nondestructive test, therefore, must be to assess the integrity of a coating and hence confirm its ability to perform the function for which it is intended. The various NDT methods that are becoming available for coatings have the potential to measure different physical properties, and each, therefore, will give a different measure of integrity. An NDT method chosen to assess a coating must be appropriate to its particular end use.

17.3 NONDESTRUCTIVE METHODS FOR EVALUATION OF SPRAY COATING PROPERTIES

NDT methods are essentially based on interactions between the component and some specific form of electromagnetic radiation, which may include

- radiated electrons
- ions

- X-rays
- eddy currents
- light
- sound

- induced magnetic fields
- thermal radiation
- microwaves
- beta back-scattering

The waves or radiation are applied externally to the test component, in such a manner that they are able to penetrate it and effect some registration of the component's surface and/or internal conditions.

Useful summaries of the progress of NDT as applied to sprayed ceramic and metallic coatings have been provided by the Dortmund University Quality Control research team of Crostack et al. [11–13]. They have been active in developing a wide range of NDT methods for coatings. Some of their more promising contributions with potential for plasma-sprayed layers are mentioned in the following sections.

In the opinions of Crostack et al. [13], conventional nondestructive testing (NDT) techniques for bulk products often have serious shortcomings when they are used, or adapted, for the testing of thin metallic or ceramic coatings. Because the test results for a coated object will be significantly influenced by the object's surface and material properties, it is important that there exist sufficiently descriminating differences between the coating material, its inclusions, "flawed" regions and porosity, etc., and the substrate material, for the test to be meaningful. Similarly, the presence of extreme roughness at either the coating surface or of planar irregularities at a coating-coating interface or at a coating-substrate interface—as well as certain complexities in the geometry and microstructural properties of the respective layers, etc.—may inhibit the effectiveness of an otherwise potentially usable NDT method. Thus, for the NDT test to be meaningful in its ability to describe and detect the presence of coating flaws, they considered it may be necessary to either [13] modify a number of the otherwise conventional techniques (e.g., eddy current pulse, acoustic emission and holographic soundfield visualisation methods) and/or to develop new techniques for specific applications (e.g., dielectric and thermal wave analysis methods).

In Germany, at least three techniques for NDT evaluation of coating thickness have been standardised. However, these have limited specific applications. They involve, respectively, magnetic methods for the measurement of nonferromagnetic layers on ferromagnetic substrates (DIN 50 981D), beta back-scattering methods (DIN 50 983), and eddy current methods for determination of electrically nonconducting layers on ferromagnetic substrates (DIN 50 984). A review of the principles, practices, and limitations of these techniques has been provided by Guhring [14].

17.3.1 DIELECTRIC MONITORING OF SPRAYED COATING POROSITY

To meet the demands for an on-line NDT porosity measurement in coatings, a new and relatively simple method has been introduced by Nelson et al. [15] for characterising the pore volume of ZrO_2–8% Y_2O_3 ceramic shroud coatings through an assessment of the coatings' dielectric properties. For this approach the coating is considered to be a two phase composite mixture, in which the ceramic represents the first (continuous) phase, and the porosity is the second phase. From previously established criteria – i.e., regarding the dielectric characteristics of the ceramic phase and the results of prior research which showed that pore size, content and distribution have a direct and measurable effect on a plasma-sprayed coating's dielectric properties – and by making on-line dielectric measurements on the G.E. (ceramic shroud) coating production, and then applying the appropriate "rule of composite mixtures" criteria that they established, Nelson et al. were able to "extract" the second phase volume, and hence monitor, characterise, and control the degree of porosity contained in the plasma-sprayed shroud coatings.

17.3.2 *IN SITU* ACOUSTIC EMISSION MONITORING OF THE PLASMA-SPRAY PROCESS

The technique of acoustic (or stress-wave) emission operates on the principle that, as strain energy is released in a solid, there is an accompanying emission of a vibrational wave that can, in effect, be detected by a transducer at the surface. For sprayed coatings, Bucklow [10] suggested the technique might only find application in strictly comparative situations, where the "noises coming out" would be compared with those emitted by a standard component of proven reliability. However, he realised there probably would be too much information in the detected signal(s), which would lead to major problems in interpretation. Following developments by Almond et al. [16,17] – who were to use AE used in association with four-point bend (i.e., destructive) testing [18] – Crostack et al. [12] found that AE had the potential to draw a correlation between some spraying parameters and acoustic emission signals emitted during the cooling of a coating.

From their work with arc-flame spraying, Crostack et al. proposed that a direct relationship between AE signals and coating characteristics might be possible on the basis of existing and sufficiently known interdependences between spraying parameters and mechanical-technological characteristics. This team demonstrated – as in Figure 17.1(a) and (b) [12] for flame spraying – that certain process variables, including arc current

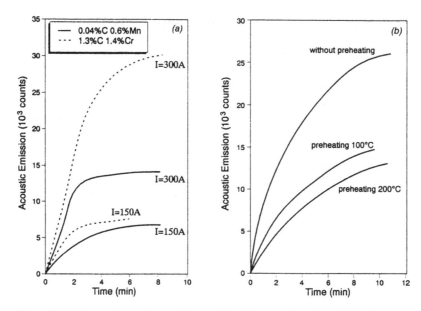

Figure 17.1 Influence of operating conditions, such as (a) current intensity and (b) substrate pre-heat temperature—on the intensity of acoustic emission (after Crostack et al., 1985).

intensity (a) and preheat temperature of the substrate (b), had a distinct effect on the sound emitted by the coating during the cooling process. For Figure 17.1(a), they considered that the total number of counts rose with increasing current intensity and suggested that increased heat introduced greater stresses into the substrate. As these stresses decreased while the substrate cooled down—i.e., by plastic deformation and microcracks—they were accompanied by higher levels of AE. With Figure 17.1(b), Crostack et al. suggested that AE was associated with variations in the respective coatings' contraction following different preheats. Because the deformation (shrinkage strain) was greater in the case of originally colder substrates, higher levels of AE resulted.

By contrast, Almond and Reiter [19] were more cautious in their predictions for acoustic emission as an NDT technique. They reported that, whilst under certain circumstances the acoustic emission technique "can reflect coating integrity, the method has serious limitations."

More recent efforts by Pacey and Stratford [20] have again stimulated speculation regarding this method of NDT for sprayed coating. Their work has attempted to relate spraying parameters to characteristic AE coating signals with moderate success. In experiments designed to simulate particle impact, they found that differences in AE signal could be related to "hard" and "soft" particle behaviour. They observed discernible differences in AE signals for plasma-sprayed coatings deposited under various arc currents, with the signals clearly different for widely different currents. However, signals also were found to vary within a run, and the early discernible differences that were seen for different sprayed materials were found to become less as spraying time increased. In addition, different spray materials behave differently with respect to varying arc current. Arithmetic means of acoustic signal distribution counts were found to provide an indication of trends in "pure" spray materials (e.g., aluminium titanate, nichrome, etc.), but results were misleading for the deposition of composites and blends of these materials. Distributions of acoustic parameters were found to be sensitive to different conditions, with rise time and amplitude being amongst the most sensitive of these parameters. Provisional results by Pacey and Stratford on some coated materials also showed "discernible differences between the characteristics of different particle morphology—spherical, fused and crushed or composite blend."

Pacey and Stratford suggested that, by correlating AE characteristics to those properties already commonly examined in conjunction with destructive testing, viz., porosity, adhesion, and cohesion, then perhaps "acoustic emission could become a rapid, nondestructive method of *in situ* quality control." They visualised the development of a "robust, possibly single-parameter monitor" unit for quality control, i.e., one possessing a high data capture and processing capability that "could also be applied to process control."

17.3.3 THERMAL WAVE INTERFEROMETRY FOR DETECTION OF ADHESION DEFECTS AND MEASUREMENT OF COATING THICKNESS

The technique of thermal wave interferometry, based on photoacoustic spectroscopy [21], was introduced by Almond and his Bath University colleagues some eight years ago [19,22–24] as a possible method for monitoring the thickness of a plasma-sprayed coating and for the detection of adhesion defects contained therein. In thermal wave interferometry testing, a periodically modulated laser light source is used to affect surface temperature modulation of the specimen. The resultant thermal waves, which are strongly damped, propagate through the coating. The waves are confined to very narrow ranges, which range over about one thermal wavelength (i.e., typically for metals some hundreds of microns at laser modulation frequencies of approximately 100 Hz). Reflections of the waves occur at coating/substrate interfaces or at flaws within the coating. Thus, when the modulated heat source is scanned over the specimen surface, these reflections effect discernible changes in the amplitude and phase of the temperature modulation on and near the surface, and thereby can provide information regarding the coating's thickness and its flaw content. A

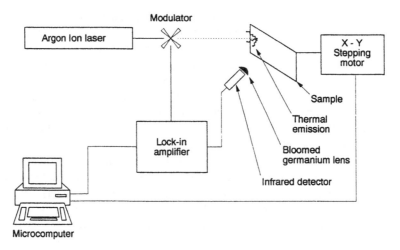

Figure 17.2 Schematic of an experimental thermal wave analysis system as developed at Bath University (after Patel and Almond, 1985).

schematic of the Patel and Almond experimental system [24] is shown in Figure 17.2.

Figure 17.3 is from subsequent experimental research conducted by Almond et al. [22]. It indicates the possibility for the thermal wave technique to be used for the measurement of plasma-sprayed coating thickness. Following the irradiation of a specially manufactured step-profile spe-

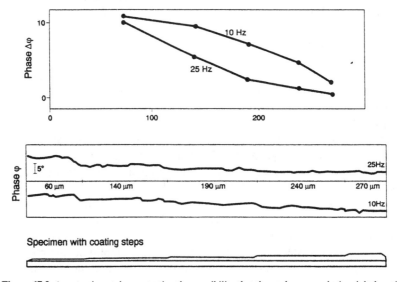

Figure 17.3 An experiment demonstrating the possibility that thermal wave analysis might be utilised to evaluate a coating's thickness (after Almond et al., 1985).

cimen—i.e., a metal substrate spray-coated with Metco grad 81 (Ni-Cr carbide) to five stepped thicknesses between 60 and 270 micron—at the respective thermal wave modulation frequencies of 10 and 25 Hz, each step (or change) in coating thickness was observed by the Bath team to be accompanied, in turn, by a change of phase ϕ. They established that the incremental sensitivity of the phase change, $\Delta\phi$, was greatest with the smaller coating thicknesses.

Suitable refinement of the test parameters would appear necessary should the technique of thermal wave analysis be used for the detection of flaws in sprayed coatings of different materials and of different thicknesses. Bennett and Patty [21] showed theoretically that the amplitude of the surface temperature oscillation will be influenced by changes in the reflectivity of the coating-substrate interface when the coating thickness is less than three-quarters of the thermal diffusion length. These workers also determined that the phase angle will be affected for coating thicknesses of up to one and a half times the thermal diffusion length. As is shown in Figure 17.4, Almond et al. confirmed that an interfacial adhesion defect (for a plasma-sprayed aluminium coating on a steel substrate) will be detectable at much higher frequencies (i.e., shorter diffusion lengths) in the phase angle measurement than in the amplitude measurement. This team [22] noted that the phase angle evaluation was found to be independent of small changes in the optical absorption and infrared emissivity of the aluminum

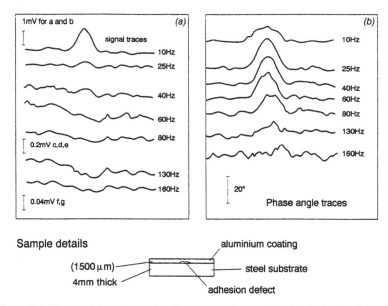

Figure 17.4 The use of thermal wave interferometry to detect an interfacial adhesion defect, by using (a) amplitude and (b) phase angle evaluation techniques (after Almond et al., 1985).

coating, thereby affording genuine subsurface information. However, it seems likely [12] that changes in the signal amplitude could be influenced by surface inhomogeneities, as well as subsurface flaws.

It would appear that thermal wave interferometry techniques are limited to the examination of samples that can be brought up to the testing equipment, and for which it is possible to achieve line-of-sight between detector and sample. However, with the recent advances that have taken place in laser technology for heat sources (e.g., gas lasers, diode lasers, etc.), in fibre optics for energy transfer (removing the requirement of line-of-sight energy transfer), and in rugged detection systems (e.g., triglycine sulphate pyroelectric detectors) for infrared radiation detection, Morris et al. [25,26] have been engaged in the development of a portable prototype system for which they envisaged the possibility for on-site testing of coatings. This team has recognised a limitation to this remote optical fibre system possibility, through the high losses inherent in the optical fibre transmission of infrared radiation. Nevertheless, they appear to express optimism in regard to this type of coating assessment procedure, a schematic of which is given in Figure 17.5.

17.3.4 CS-IMPULSE EDDY CURRENT TECHNIQUES FOR DETECTION OF CRACKS

The use of the CS-Impulse (also known as Controlled Signal Pulsed) Eddy Current Technique for the testing of thermally sprayed CoCrAlY and

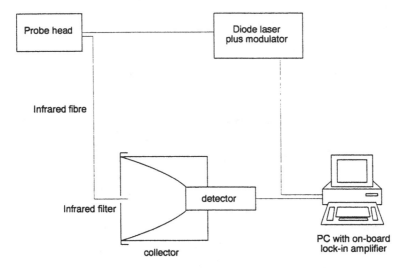

Figure 17.5 Schematic block diagram of a transportable photothermal radiometry system (after Morris et al., 1989).

steel coatings has been developed by Crostack and Jahnel [27]. It is claimed [13,28] that eddy current testing of electrically conducting material coatings is capable of "the detection of manufacturing flaws such as cracks within the coatings, and hidden bonding flaws between coating and substrate down to a diameter of 3 mm." It is additionally claimed that "the early rejection of defective components is possible during production." A disadvantage of testing conducting coatings by conventional single-coil mono-frequency eddy current techniques is the problem of interpreting the measured values. This is particularly so when useful signals are superimposed upon extraneous (noise) signals, and when a clear identification of, and discrimination between, the possible influences on the two characteristic signal parameters (viz., amplitude and phase) involved in the technique is not achievable.

Some of the ambiguities associated with the mono-frequency eddy current technique when it is used to test sprayed coatings have been described by Crostack et al. [28]. Signals measured for, and attributable to, bonding flaws, localised porosity, or cracks may be overlaid by the signals representing the manufacture-induced variations in coating thickness or by those associated with the background porosity. In addition, there may be signal perturbation problems due to transducer inclination, geometric limitations (including that of the substrate), each of which may make signal interpretation even more difficult. From the work of this Dortmund team, it was concluded that mono-frequency testing results were ambiguous, and in order to effect an improvement, they considered it was necessary to utilise an appropriate combination of frequencies, to suppress the disturbing frequencies, and thereby correlate individual signals with specific anomalies of defects associated with a coating. They determined that the excitation of a useful range of frequencies could be achieved by pulse techniques and that a single, short test pulse can be used to excite a complete frequency band, and the pulse responses could be evaluated on-line—in the time domain.

According to Crostack et al. [29] and Nehring [30], when a digital transmitter unit is utilised for the CS (Controlled Signal) Pulsed Eddy Current technique, this allows the time domain, and hence the spectral content of the excitation pulse, to be defined. Also, by the incorporation of a facility to suitably match the frequency band of a CS-Pulse to a specific penetration range (which also readily accommodates the characteristic eddy current behavioural variations associated with differences in material composition), these workers claim to have devised an improved digitised eddy current signal processing technique that permits "rapid and easily performed examination" of sprayed coatings. An advantage also appears to be the ability to optimise the transmitting impulse by using a cross-correlation technique, with the response signal of a precalibrated crack (or

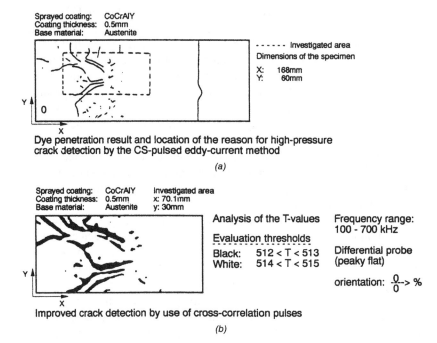

Sprayed coating: CoCrAlY
Coating thickness: 0.5mm
Base material: Austenite

- - - - - - Investigated area
Dimensions of the specimen

X: 168mm
Y: 60mm

Dye penetration result and location of the reason for high-pressure
crack detection by the CS-pulsed eddy-current method

(a)

Sprayed coating: CoCrAlY   Investigated area
Coating thickness: 0.5mm   x: 70.1mm
Base material: Austenite   y: 30mm

Analysis of the T-values

Evaluation thresholds

Black: 512 < T < 513
White: 514 < T < 515

Frequency range:
100 - 700 kHz

Differential probe
(peaky flat)

orientation: $\frac{0}{0}$ -> %

Improved crack detection by use of cross-correlation pulses

(b)

Figure 17.6 Presence of manufacturing flaws in a CoCrAlY plasma-sprayed coating as revealed (a) on the complete specimen area by a die penetrant technique and (b) on a reduced (selected) area of the specimen by a high-resolution crack detection CS-pulsed eddy current method utilising cross-correlation pulses (after Crostack et al., 1989).

other type of defined flaw) chosen as the reference. The calibration can be fully automated [30].

A detailed introduction to, and possible application suitability for, this inspection technique in the detection of manufacturing flaws in coated products has been given by Crostack et al. [28]. Its use for the detection of flaws (pre-identified by dye penetrant) in a CoCrAlY plasma-sprayed coating on an austenitic substrate is illustrated in Figure 17.6 [13].

17.3.5 HOLOGRAPHIC SOUNDFIELD IMAGING

Fischer, in 1982 [31], recognised the possibility of using holographic sound imaging for the investigation of the flaws in plasma-sprayed metallic and ceramic coatings. This method is based on a combination of holographic interferometry and conventional ultrasound testing techniques to "visualise" sound waves transmitted into the sprayed component's surface.

In the imaged soundfield, close-to-the-surface flaws, such as substrate-coating bonding flaws and cracks in the coating, are detected by their interactions with the induced sound waves, i.e., through the localised perturbations that the flaws cause to the soundfield image, the size and shape of these defects can be displayed [11,32].

After extensive research on the Dortmund University equipment [33], a mobile holographic camera (SFK) was developed. This equipment is reported [34,35] as being capable of on-site contact-free holographic-interferometric imaging of surfaces of industrial products. From this Dortmund research, "outside the laboratory" applications were largely met by the use of a frequency-stabilised pulsed (ruby) laser—with a pulse energy of 0.5 joule—installed in the camera head. In addition, Crostack and his team were satisfied that, following suitable modifications, the system was made virtually vibration-free with respect to external shock and vibrational disturbances, so that it could be used for the examination of moving objects in industrial (low-pressure, plasma-spraying) situations.

It would appear that this type of equipment—when it has been suitably calibrated with regard to known flaw parameters, and when it is operated within the predetermined frequency limits of its "flaw-detection" capability—already offers some potential as a quality control method for sprayed coatings. However, it would also seem that its current application is limited to bonding flaws down to a size <30% of, and to depth regions of <60% of the ultrasonic wavelength [13,35]. In addition, cracks in a sprayed coating are most sensitively detected if the test frequency (which is often between 0.5 and 1.0 MHz) corresponds to a wavelength smaller than four times the coating thickness. As Figure 17.7 [35] shows, while bonding flaws of dimensions between 4.5 and 10 mm are clearly detected at a test frequency of 0.5 MHz in a CoCrAlY coating 0.5-mm thick, the unambiguous detection of smaller sized flaws in a similarly compositioned coating of 0.25-mm thickness was only achievable by an increase in testing frequency to 1.0 MHz.

17.3.6 POSSIBILITY FOR OTHER NONDESTRUCTIVE TESTING METHODS

Among other NDT methods that may find application for sprayed coatings are those involving fast infrared scanning and photothermal techniques (see, e.g., Hartikainen [36–38] and Cielo and Dallaire [39]). The application of infrared thermography for the testing of coatings received serious attention [40], but so far no practical application seems to have developed from this research.

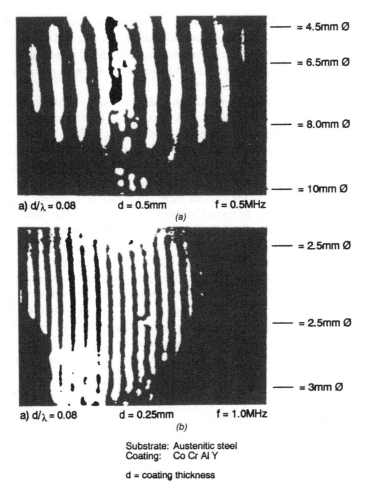

Figure 17.7 Use of holographic soundfield imaging techniques to reveal coating flaws in CoCrAlY coatings with the respective thicknesses of 0.5 mm (a) and 0.25 mm (b) on stainless steel substrates (after Crostack and Pohl, 1988).

17.4 WHICH NDT METHOD TO USE FOR PLASMA-SPRAYED COATINGS

Of the five above-mentioned promising NDT methods for assessing a specific coating "quality," only dielectric monitoring of porosity in ceramic gas path seal shrouds has been introduced into a production-line situation. If another NDT method is also to find application in a commercial spraying operation, then it will be necessary for that method to be capable of interpreting the coating's integrity as demanded by the product specification, perhaps should be transportable, and preferably be capable of being incorporated in an overall control system. Perhaps such a goal will be achieved through the rapid developments being pursued in transportable holographic soundfield imaging, or from the energetic research being conducted into the controlled signal pulsed eddy current testing. However, any decision to develop commercial evaluation equipment from this type of research and any chances of it succeeding commercially will of course, ultimately, be based on economic considerations. It must be shown that the control over quality that might be obtained from the apparatus significantly outweighs the various contained development, purchase, and installation costs, which are associated with its commissioning as a production capital item.

While significant advances towards nondestructive methods for assessment of spray coating parameters may have been made, there still remains a considerable reliance on destructive methods to confirm the quality of samples of contemporary coated products. In addition, it would appear that the achievement of effective quality assurance for today's coatings does not rely on the sophisticated testing procedures that might be utilised, whether these be destructive or nondestructive, but rather more on the strict control that is being maintained at each stage of the processing cycle.

An important aspect of any quality control program is the ability to identify and rectify unobserved processing inadequacies as early as possible. A failure to observe one or more of a number of process inadequacies, such as inappropriate design, faulty powder, insufficient surface preparation, incorrect spraying pattern, etc., which can separately contribute to the production of an inferior quality coating (such as one possessing low adhesive strength), may prove disastrous. Test criteria need to be examined very carefully. It is important if and when a test (destructive or otherwise) is to be introduced, that it should be consistent with the component's application, and that inspection is, in fact, a verification of the observance of good practice. A situation should not be allowed to develop where a sprayed component can successfully meet specification tolerances and pass an apparently appropriate NDT examination, only to have that product's inherent inadequacy revealed when, at some time later, it has failed to withstand the rigours of its in-service application. Obviously, as Smart correctly expressed it some fifteen years ago [9], "Poor quality cannot be tested out! Good quality MUST be built in!"

17.5 ACKNOWLEDGEMENTS

The preparation of this review was made possible through funding provided by the I. R. and D. Board's New Materials Generic Technology

Grant Number 15007 for research into "Electric Arc Processing of Surface Treatment." This support is gratefully acknowledged.

The encouragement provided to the author by CSIRO colleagues, notably Dr. N. Gane for his useful discussions, suggestions, and critical reading of the text, is most appreciated.

17.6 REFERENCES

1 Doolette, A. G., Oppenlander, W. T. and Ramakrishnan, S. "Electric Arc Generating Equipment—Allows Rapid Arc Power Variation and Power Distribution," Australian Patent No. WO9003095, 1990.

2 Houghton, M. E. "Automation and Control of Plasma Spray Processes, a Review of the Literature," CSIRO Division of Manufacturing Technology Technical Report No. MTM 187, 178 pp., 1990.

3 Berndt, C. C. and Ostojic, P. "Strength Testing of Plasma Sprayed Coatings," *Proc. of ATTAC '88,* Osaka, pp. 191–198, 1988.

4 DIN 50160. "Ermittlung der Haft-Zugfestigkeit im Stirnzugversuch" (Determining of Adhesive Tensile Strength in a Front-on Tensile Test), Berlin, 1981.

5 ASTM 633-79. "Standard Method of Test for Adhesion of Cohesive Strength of Flame-Sprayed Coatings," Part 17, *Annual Book of ASTM Standards,* 1982.

6 AFNOR NF A91-202-79. "Characteristics and Methods of Test for Metal Spraying," Paris, 1979.

7 JIS H 8666-80. "Thermal-Sprayed Ceramic Coatings," Tokyo, 1981.

8 SAE AMS 24378. "Coating, Plasma Sprayed Deposition," Warrendale, PA, 1986.

9 Smart, R. F. "Quality Control of Sprayed Coatings," *Proc. of 8th Intl. Conf. on Thermal Spraying, ITSC 76,* Miami, pp. 195–202, 1976.

10 Bucklow, I. A. "Possible Methods of Non-Destructive Testing of Sprayed Coatings," *Materials in Engineering,* 2(3):141–148, 1981.

11 Crostack, H.-A., Kruger, A., Fischer, W. R. and Steffens, H.-D. "Nondestructive Testing of Thermally Sprayed Coatings by Using Optical Holography to Receive Ultrasonic Waves," *DVS Berichte,* 16:28–30, 1983.

12 Crostack, H.-A., Steffens, H.-D., Kruger, A. and Pohl, K.-J. "Non-Destructive Testing of Sprayed Ceramic and Metallic Coatings," *Proc. of Conf. on Advanced Materials R and D for Transport, Ceramic Coatings for Heat Engines,* Strasbourg, Nov. 26–28, 1985, Les Editions de Physique, pp. 203–220, 1986.

13 Crostack, H.-A., Jahnel, W., Meyer, E. H. and Pohl, K.-J. "Recent Developments in Non-Destructive Testing of Coated Components," *Thin Solid Films,* 181:295–306, 1989.

14 Guhring, W. H. "The Latest Techniques and Instruments for Measuring the Thickness of Metallic Coatings," *British Journal of NDT* (Sept.):251–258, 1982 and (Nov.):321–326, 1982.

15 Nelson, W. A., Grossman, T. R. and Gehrhardt, R. "Nondestructive Characterisation of Porosity in Ceramic Shroud Coatings," paper to *3rd National Thermal Spray Conf.,* Long Beach, CA, May 20–25, 1990.

16 Almond, D. P., Cox, R. L. and Reiter, H. "Acoustic Properties of Plasma Sprayed Coatings and Their Application to Non-Destructive Evaluation," *Thin Solid Films,* 83:311–324, 1981.

17 Almond, D. P. "An Evaluation of the Suitability of Ultrasonic Techniques for the Testing of Thermally Sprayed Coatings," *Surfacing Journal,* 13(3):50–55, 1982.

18 Almond, D. P., Moghisi, M. and Reiter, H. "The Acoustic Emission Testing of Plasma Sprayed Coatings," *Thin Solid Films,* 108:439–447, 1983.

19 Almond, D. P. and Reiter, H. "Novel Ways of Looking at Thermal Spray Coatings," *Surfacing Journal,* 16(1):4–11, 1985.

20 Pacey, R. A. and Stratford, V. "In-situ Acoustic Emission Monitoring of the Plasma Spraying Process," *Proc. of 1st Plasma-Technik Symposium,* Lucerne, 1:135–144, 1988.

21 Bennett, C. A., Jr. and Patty, R. R. "Thermal Wave Interferometry: A Potential Application of Photoacoustic Effect," *Applied Optics,* 21:49–54, 1982.

22 Almond, D. P., Patel, P. M., Pickup, I. M. and Reiter, H. "An Evaluation of the Suitability of Thermal Wave Interferometry for the Testing of Plasma Sprayed Coatings," *NDT International,* 18:17–23, 1985.

23 Patel, P. M., Almond, D. P. and Reiter, H. "Photothermal Testing of Coating Surfaces," Paper to *Conf. on Acoustic Emission and Photoacoustic Spectroscopy, Proc. Inst. of Acoustics,* London, 1983.

24 Patel, P. M. and Almond, D. P. "Thermal Wave Testing of Plasma Sprayed Coatings and a Comparison of the Effects of Coating Microstructure on the Propagation of Thermal and Ultrasonic Waves," *J. Mat. Sci.,* 18:955–960, 1985.

25 Morris, J. D., Patel, P. M., Almond, D. P. and Reiter, H. "Thermal Wave Characterisation of Translucent Ceramic Coatings," *Proc. of 5th Intl. Meeting on Photoacoustic and Photothermal Phenomena,* Heidelberg, pp. 427–429, 1987.

26 Morris, J. D., Almond, D. P., Patel, P. M. and Reiter, H. "Developments in Thermal Wave Non-Destructive Testing Systems for Thermal Spray Coatings," *Proc. of 12th Intl. Conf. on Thermal Spraying, ITSC '89,* London, 1:99-1–99-6, 1989.

27 Crostack, H.-A. and Jahnel, W. "Zerstorungsfreie Gutebestimmung von thermischen gespritzten MCrAlY-Schutzschichten mit dem CSImpulswirbelstromverfahren," *DVS Report,* no. 1.195; (also Arbeitsgemeinschaft Industrieller Forschungsvereinigen Report, no. 6511), Dortmund, 1988.

28 Crostack, H.-A., Jahnel, W., Kohn, J. and Polaud, B. "Non-Destructive Quality Evaluation of Thermally Sprayed MCrAlY Protective Coatings with CS-Pulsed Eddy Current Method," *Proc. of 12th Intl. Conf. on Thermal Spraying, ITSC '89,* London, 1:55-1–55-13, 1989.

29 Crostack, H.-A., Nehring, J. and Oppermann, W. "Einsatz von Korrelationsimpulsen (CS-Technik) zur Wirbelstromprufung," *Materialprufung,* 26:16–20, 1984.

30 Nehring, J. "Untersuchungen zum Einsatz problemangepasster Prufimpulse (CS-Technik) bei der Wirbelstromprufung," Dissertation, University of Dortmund, 1987.

31 Fischer, W. R. "Beitrag zur Weiterentwicklung der holographischen Interferometrie zur zer storungsfreien Prufung thermisch gespritzter Schichten," *Fortschritt-Berichte der VDI-Zeitschriften,* series 5, no. 66, VDI-Verlag, Dusseldorf, 1982.

32 Crostack, H.-A. and Fischer, W. R. "Einsatz von Ultraschall in Verbindung mit holographischer Interferometrie zur zerstorungsfreien Prufung thermisch gespritzter Schichten," *Materialprufung,* 24(2):49–54, 1982.

33 Crostack, H.-A., Krüger, A. and Pohl, K.-J. "Zerstorungsfreie Qualitatssicherung von Beschictungen," *VDI-Berichte,* no. 624, 1986.

34 Krüger, A. "Beitrag zur Prufung oberflachennaher Bauteilebereiche mittels

Ultraschall durch Einsatz eines mobilen holographischen Schallabbildungssystem," *VDI-Fortschritt-Berichte,* series 5, no. 122, VDI-Verlag, Dusseldorf, 1987.

35 Crostack, H.-A. and Pohl, K.-J. "Quality Control of Plasma Sprayed Coatings by Means of Holographic Sound Field Imaging," *Proc. of 1st Plasma-Technik Symposium,* Lucerne, 2:253–265, 1988.

36 Hartikainen, J. "A Fast Infrared Scanning Technique for Non-Destructive Testing," *Review of Scientific Instruments,* 60:670–672, 1989.

37 Hartikainen, J. "Inspection of Plasma-Sprayed Coatings Using Fast Infrared Scanning Techniques," *Review of Scientific Instruments,* 60:1334–1337, 1989.

38 Hartikainen, J. "Fast Photothermal Measurement System for Inspection of Weak Adhesion Defects," *Applied Physics Letters,* 55:1188–1190, 1989.

39 Cielo, P. and Dallaire, S. "Optothermal NDE of Thermal Barrier Coatings," paper at *ASM's Intl. Conf. on Surface Modifications and Coatings,* Toronto, 10 pp, 1985.

40 Pawlowski, L., Lombard, D., Mahlia, A., Martin, C. and Fauchais, P. "Thermal Diffusivity of Arc Plasma Sprayed Zirconia Coatings," *High Temperatures, High Pressures,* 16:347–359, 1984.

CHAPTER 18

Fatigue of Surface-Hardened Gears

T. P. WILKS[1]
G. P. CAVALLARO[1]
K. N. STRAFFORD[1]

18.1 INTRODUCTION

18.1.1 BACKGROUND AND RATIONALE

SIGNIFICANT technical advances have been made in the past decade in steel making, heat treatment, and gear design. In heat treatment, plasma carburizing [1] and plasma nitriding [2–4] have led to significant quality improvements and energy savings. Progressive tooth-by-tooth and spin induction hardening techniques are well established and have been combined with automated precision forging techniques for the production of near net shape gears with improved mechanical "fibre."

Gears used in power transmission are often subject to very severe operating loads, and in order to assure their reliability, they are designed and manufactured to strict quality assurance standards. Transmission gears are normally spur or helical gears, with bevel gears in rear axle assemblies.

Spur gears are usually cut from bar stock (hobbed) before being carburised, followed by a quench and temper heat treatment and glass bead peening. Another option is to use tooth-by-tooth induction hardening in place of the quench and temper, depending upon the design stresees. Heat treatment must be carried out with minimum distortion [5] and be operated at relatively low unit cost. There are many competing heat treatment processes, and it is usual to study them in two categories:

(1) *Thermochemical:* carburising, carbonitriding, nitriding, nitrocarburising, boriding (where chemical is added)

---

[1]Metallurgy Department, University of South Australia, Adelaide, SA 5095, Australia.

(2) *Thermal:* quench and temper, induction harden, flame harden, laser harden [6–9], electron beam harden [9]

Figure 18.1 shows an illustration of the types and sources of gear blanks, the choice being a compromise between cost and performance. The cheapest route is to machine the gear from rolled bar; however, gears subject to higher loads may require a forging operation. Heat treatment processes must be selected to prevent or minimise distortion whilst achieving the desired mechanical properties. Figure 18.2 is a schematic illustration showing some of the competing methods of heat treatment used in the manufacture of transmission parts.

In order to optimise the performance of gear components by careful selection of materials, processing route, and heat treatment, it is necessary to fully understand the mechanism(s) by which failure of the gears occurs and relate these to material and process variables.

The present research study examined the fatigue of surface-hardened gears and its prevention through optimised manufacturing technology. A testing program was devised to examine the influence of gear design, material processing, heat treatment, and steel cleanliness, as well as surface engineering considerations such as surface-hardening practices, surface finish, and residual stresses. The aim of the program was to examine the influence of the above parameters on the resistance to fatigue of spur gears manufactured from EN39B steel. Studies focussed on the fatigue

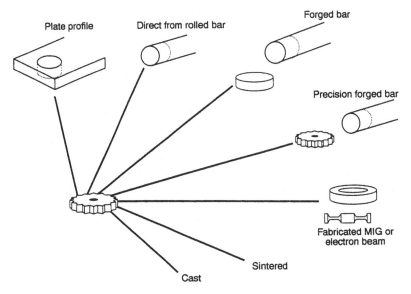

Figure 18.1 Gear blank sources.

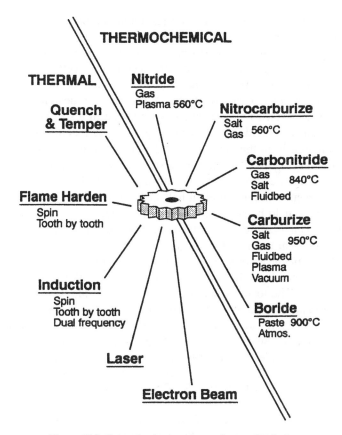

Figure 18.2 Competing heat treatments for gear hardening.

fracture process [10], and extensive fractography was carried out to identify the crack path location, initiation, and growth relative to the case position [11].

18.2 EXPERIMENTAL

18.2.1 GEAR FATIGUE TESTING (SINGLE TOOTH BENDING)

To study the bending fatigue characteristics of gear teeth during single tooth bending tests, the experimental setup involves uniaxially loading single gear teeth at the correct pressure angle to simulate deformation and loading conditions encountered during service. Fatigue cracks are initiated at the root of the gear tooth, which coincides with the area of high stress concentration, and crack propagation results in complete fracture, which separates the tooth from the main gear wheel.

The apparatus required to perform single tooth bending fatigue tests was designed to simulate deformation experienced by the teeth of a spur gear during service. Provision had to be made in the design of the testing rig to facilitate testing of consecutive teeth. A test rig was designed and constructed, which consisted of a spur gear mounted in a specially designed holder, which was fitted to an Instron fatigue testing machine (100 KN) to actuate the punch on the gear tooth at the correct pressure angle.

This design (Figure 18.3) relies on a specially constructed hub that consists of a spline designed to mesh with the centre of the gear, which is fixed, to prevent rotation, by eight half-inch UNF bolts. These bolts were designed to be held by slots that allow a fine rotation of the gear to align it with the punch before final tightening. The interior spline was indexed to the outside gear teeth so that the gear could be extracted from its holder, rotated by a suitable amount, and re-inserted to test the adjacent tooth. The new position maintains the same tooth orientation, with respect to the actuating punch. The load applied took the form of a sine wave, cycling between zero and maximum load at a frequency of 40 Hz, allowing fatigue lives up to 10^7 cycles to be achieved in a reasonably short time scale.

Test gears were twenty-four tooth, 5 DP (diameter pitch), with a 20° pressure angle, made from EN39B steel, cut from bar stock, carburised, quench and tempered, and glass bead peened. A typical analysis for EN39B steel is 0.15% C, 0.25% Si, 0.5% Mn, 4.2% Ni, 1.2% Cr, and 0.2% Mo.

Results were presented in the form of stress versus cycles (S-N) data in order to define a fatigue limit for the material/specimen geometry, and examination of fracture surfaces was carried out using optical and scanning electron microscopes (SEM), the latter being fitted with a facility for chemical analysis.

The structure of the steel (not carburised) is a tempered martensite with small amounts of retained austenite and a fine dispersion of small carbides. The minimum specified tensile strength for this steel is 1310 MPa, with Brinell hardness not exceeding 269 HB (tempered condition).

18.2.2 FRACTURE SURFACE EXAMINATION

Examination of the fracture surfaces of failed gear teeth, using optical and scanning electron microscopes (SEM), showed that each of the teeth failed due to the initiation and growth of a single fatigue crack (Figure 18.4). Each fatigue crack initiated at subsurface, elongated manganese sulphide (MnS) inclusions positioned 150–300 μm below the surface of the gear tooth root in the area of high stress concentration (Table 18.1). The in-

(a)

(b)

Figure 18.3 Showing (a) bending fatigue apparatus and (b) details of test gear and indexing hub.

335

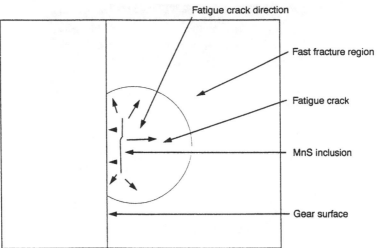

Figure 18.4 Fatigue crack initiation site in a carburised EN39B steel gear tooth (magnification × 100).

TABLE 18.1. Results of Bending Fatigue Tests Carried out on Individual Gear Teeth.

Tooth Number	a (mm)	b (mm)	c (mm)	d (mm)	e (μm)	Stress (MPa) Load (kN)	Cycles (N)
1						Overload	Overload
2	0.5	0.65	0.3	0.2	66	1270, 35	1.05×10^6
3	0.5	0.75	0.2	4.5	50	1090, 30	2.66×10^6
4						540, 15	No failure
5	0.4	0.5	0.2	13.5	98	1380, 38	8.1×10^5
6	0.6	1.0	0.2	2	42	910, 25	2.23×10^7
7						Overload	Overload
8	0.55	0.9	0.25	13	80	1020, 28	7.12×10^6
9	0.45	0.8	0.2	12	41	1200, 33	9.6×10^5
10	0.55	0.95	0.25	14	36	1130, 31	4.47×10^6
11	0.45	0.7	0.2	5	145	1130, 31	6.0×10^5
12	0.45	0.7	0.15	0.9	100	1130, 31	7.55×10^5

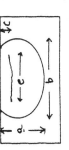

d = Distance of initiation site from edge of tooth.
e = Apparent length of inclusion (visible portion).

clusions were identified using an energy dispersive spectrometer (EDS) attached to the SEM, and the X-ray spectrum produced (Figure 18.5) confirmed the presence of MnS. The cracks grew to depths of 400–600 μm below the surface (see Table 18.1) before fast fracture occurred in a mixed intergranular/transgranular mode (Figure 18.6). The fatigue cracks lay within the carburised layer, which extended to a depth of approximately 1.4 mm, deduced from metallographic examination and microhardness measurements (Figure 18.7). There was no evidence of fatigue crack initiation from any inclusions on, or less than 150 μm from, the surface. The gears were glass bead peened prior to fatigue testing, which would introduce compressive residual stresses into the near-surface material.

Compressive residual stresses associated with carburised steel surfaces that have been shot peened have been shown [12,13] to extend to a depth of approximately 150 μm below the surface. The existence of compressive residual stresses in the carburised layer of the gear would account for the lack of initiation of fatigue cracks from inclusions lying close to the surface of the gear. The inclusions on or near the surface are surrounded by compressed material, which inhibits fatigue crack initiation; however, as the compressive residual stress reduces, or becomes tensile, away from the surface, the stress concentrating effect of the inclusions may be sufficient to locally deform the metal matrix and initiate fatigue cracks.

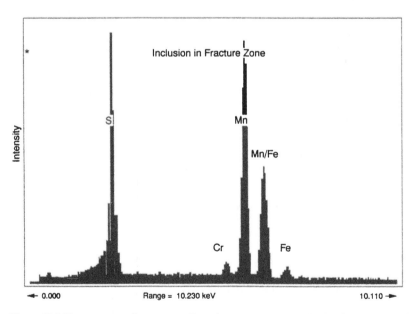

Figure 18.5 X-ray spectrum from energy dispersive spectrometer (EDS) showing presence of manganese sulphide in the form of inclusions.

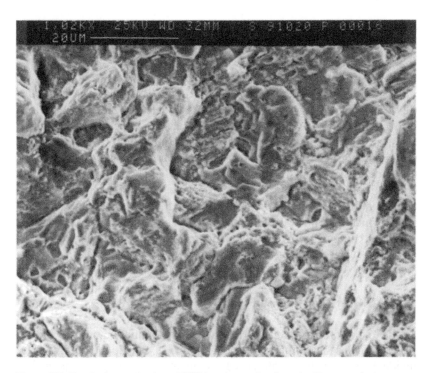

Figure 18.6 Fast fracture region in an EN39B steel gear, showing mixed intergranular/transgranular cracking in the carburised case (magnification × 1220).

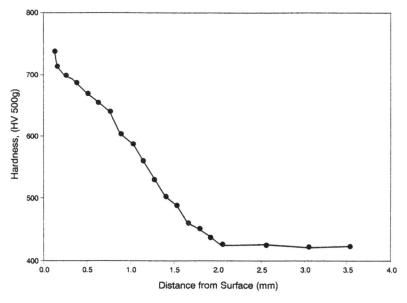

Figure 18.7 Hardness distribution of EN39B gear tooth.

339

18.3 FATIGUE DATA

Fatigue data, presented as stress (maximum) versus cycles to failure (SN), is shown in Figure 18.8. The results to date indicate a fatigue limit below 900 MPa, and for the gear teeth that failed, there was considerable variation in fatigue lives. From Figure 18.8, fatigue life (N) is a function of the maximum applied stress (σ_{max}) on the gear teeth, with increased fatigue life associated with reduced stress levels. The initiation of fatigue cracks from inclusions in martensitic steels has been shown to have a significant influence on fatigue life [14] due to the size and shape of inclusions, which determine their stress concentrating effect. The present case is further complicated by the orientation of the elongated "stringer" inclusions and surface residual stresses.

18.4 FATIGUE LIFE OF GEAR TEETH

Due to the complex nature of the fatigue crack initiation process and the many variables that influence the fatigue process, the fatigue life (N) of individual gear teeth is influenced by many interacting events and/or

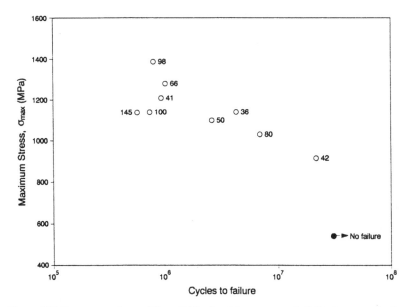

Figure 18.8 Stress vs. cycles to failure (S-N), for fatigue tests on individual gear teeth (frequency = 40 Hz) [initiating inclusion size (length in μm) shown next to data points].

phenomena. The initiation of fatigue cracks in the carburised case depends upon

(1) Size of inclusions
(2) Location of inclusions
(3) Orientation of inclusions
(4) Applied stress
(5) Stress concentrating effect due to gear geometry
(6) Stress concentrating effect due to inclusion
(7) Surface residual stresses (magnitude and extent)
(8) Mechanical properties of metal matrix surrounding inclusions

The location of the inclusions (point 2) refers to their position with respect to gear geometry and, also, the depth of the inclusions below the surface. The stress concentrating effect of the inclusion (point 6) will depend upon (a) the shape of the inclusion, (b) the hardness of the inclusions with respect to the metal matrix, and (c) the bond between inclusion and metal matrix. All of the MnS inclusions were elongated "stringer" inclusions, which had been flattened during the forging operation to produce the bar stock. This means that all of the inclusions lie approximately in the same orientation (along the original bar length) and are therefore elongated across the width of the gear. MnS inclusions are relatively soft, compared to the steel matrix, and would not by themselves exert significant localized stresses in the vicinity of the inclusion site.

The ability to influence points 1, 2, 3, and 6 is limited because they are inherent in the steel from which the gears are manufactured. The applied stress (point 4) is usually determined by service requirements, and gear geometry (point 5) is usually dictated by design criteria and specification requirements.

The purpose of this research was to look at the metallurgical aspects, both material and processing, which determine the performance of gears in service, and in this respect there is sufficient scope to influence material (point 8) and surface properties (point 7).

The mechanical properties of the carburised gear surface will depend upon previous mechanical, thermochemical, and thermal treatments. Forging of the original bar stock, gear cutting, carburising, heat treatment (quench and temper), and glass bead peening operations determines the final mechanical properties of the gear steel, especially in near-surface material. Surface hardness is strongly influenced by the carburising treatment and the subsequent quench and temper heat treatment. Hardenability is a function of the carbon level, as well as alloying elements, and the austenitising and quenching treatments control the transformation to martensite and the amount of any retained austenite. Subsequent tempering treatment determines the hardness of the case-carburised layer, which influences both the initiation of fatigue cracks and their propagation. Fatigue cracks grow to a critical size, which is a function of the stress concentrating effects, the applied stress, and the fracture toughness of the material, before fast fracture occurs, resulting in the failure of the section.

Efforts to improve the fatigue resistance of the steel must concentrate on avoiding the initiation of fatigue cracks, initiation accounting for approximately 95% of the fatigue process, with the remaining 5% concerned with crack propagation.

From the results to date, glass bead peening appears to inhibit fatigue crack initiation at the gear surface, and instead, the fatigue cracks initiate at subsurface inclusions. Tests are now under way on gear teeth that have not received glass peening treatment to determine the crack-initiating mechanism and associated fatigue life. This will establish whether the glass peening operation is necessary, in terms of improving fatigue resistance and, if so, will establish its effectiveness and potential for further improvements. It may be possible to increase the notch toughness of the subsurface material, to make crack initiation from subsurface inclusions more difficult, by altering the tempering temperature. Ideally, this may be achieved without significantly influencing the wear characteristics or fatigue resistance of the surface material, as the latter relies upon the glass bead peening process to impart fatigue resistance.

18.5 FATIGUE CRACK INITIATION FROM INCLUSIONS

To examine the influence of inclusion size and location on their ability to initiate fatigue cracks, and thus determine fatigue life, the length of the crack-initiating inclusions (dimension e in Table 18.1), as measured from fracture surface examination, was plotted against total fatigue life (N), together with the depth of each inclusion below the surface of the gear (Figure 18.9). From the results, there was no evidence that the depth at which initiating inclusions lay below the surface significantly influenced fatigue life. The size (length) of the fatigue crack-initiating inclusions did influence fatigue life, with longer inclusions (100–150 μm in length) resulting in shorter fatigue lives, compared to inclusions less than 100 μm. Three gear teeth tested at a maximum stress of 1130 MPa produced fatigue lives of 6×10^5, 7.55×10^5, and 4.47×10^6, due to fatigue cracks initiated at inclusions which were 145-, 100-, and 36-μm long, respectively (Figure 18.10).

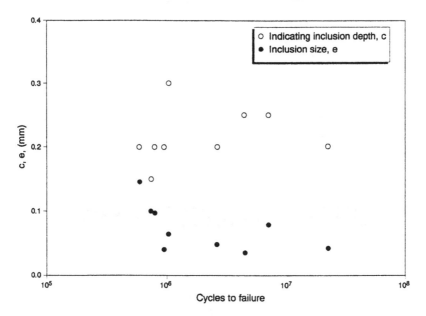

Figure 18.9 Dimension (length) of fatigue crack initiating inclusion (*e*) versus cycles to failure (*N*). Depth of each inclusion below the surface (*c*) is also shown.

18.6 FUTURE WORK

The fatigue testing program will now concentrate on the following aspects

(1) Testing of gear teeth that have not received the glass bead peening treatment (This will establish the influence and effectiveness of the peening process on the fatigue behaviour of the steel.)

(2) Altering the tempering temperature, which may alter the notch toughness of the subsurface material adjacent to the initiating inclusions and, thus, influence fatigue resistance.

18.7 CONCLUSIONS

(1) Fatigue cracks were initiated at subsurface manganese sulphide inclusions positioned 150–300 μm below the gear surface.

(2) Fatigue cracks were initiated from elongated "stringer" inclusions 36–145-μm long.

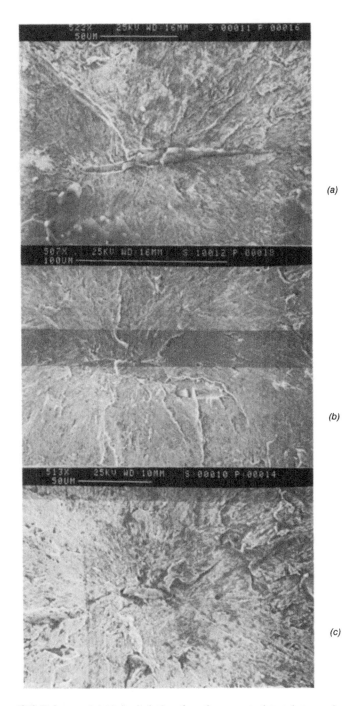

Figure 18.10 Fatigue crack initiating inclusions from three gear teeth tested at a maximum stress of 1130 MPa: (a) N $= 6 \times 10^5$ inclusion size, $e = 145$ μm; (b) $N = 7.55 \times 10^5$ inclusion size, $e = 100$ μm; (c) $N = 4.47 \times 10^6$ inclusion size, $e = 36$ μm.

(3) The initiation and propagation of fatigue cracks occurred within the surface carburised case.

(4) From the results to date, the fatigue limit for the EN39B gear steel is below 900 MPa.

(5) There was considerable variation in the fatigue lives of individual gear teeth, and this was found to be a function of the maximum applied stress and the size of the fatigue crack–initiating inclusions.

(6) The lack of fatigue cracks initiated from inclusions that lie within 150 μm of the surface is consistent with the presence of compressive residual stresses, due to the glass bead peening process, which typically has been shown to influence mechanical properties up to a depth of approximately 150 μm.

18.8 ACKNOWLEDGEMENTS

The authors would like to acknowledge Birrana Engineering Ltd for their involvement and considerable assistance with this research program. They would also like to express gratitude to Len Green, Dept. of Metallurgy at the University of South Australia, for valuable discussions and assistance.

18.9 REFERENCES

1 Jacobs, M. *Surface Eng.*, 1(2), 1985.

2 Guthrie, A. M. "Developments in Heat Treatment," *Metals and Materials* (Jan.):30, 1987.

3 Dearnley, P. A. and Bell, T. "Plasma Ion Processing of Steels," *Surface Eng. Technology 90*, Sydney, 3–4 Sept., 1990.

4 Week, M. and Schlotermann, K. "Plasma Nitriding to Enhance Gear Properties," *Metallurgia*, 8:328–332, 1984.

5 Devenny, D. "Multiflow Pressure Quenching for Distortion Free Hardening," in *Metals and Materials*, p. 88, 1990.

6 Eckersley, J. S. *Laser Applications in Metal Surface Hardening: Advances in Surface Treatments—Technology, Applications, Effects, Vol. 1*, A. Niku-Lari, ed., Pergamon Press, pp. 211–231, 1984.

7 Zenker, R. and Zenker, U. "Laser Beam Hardening of a Nitrocarburised Steel Containing 0.5% C and 1% Cr," *Surface Eng.* 5(1):45, 1989.

8 Steen, W. M. and Weerasinghe, V. M. " The Laser's Other Role—Recent Developments in Surface Coatings and Modification Processes," *Seminar at Institute of Mech. Engineers*, 1985.

9 Hick, A. J. "Rapid Surface Heat Treatment—A Review of Laser and Electron Beam Hardening," *Heat Treatment of Metals*, pp. 3–11, 1983.

10 Apple, C. A. and Krauss, G. "Microcracking and Fatigue in a Carburized Steel," *Metallurgical Transactions*, 4:1195–1200, 1973.

11 Averbach, B. L., Bingzhe, L., Pearson, P. K., Fairchild, R. E. and Bamberger, E. N. "Fatigue Crack Propogation in Carburized X-2M Steel," *Metallurgical Transactions*, 16A:1267–1271, 1985.

12 Kikuchi, M., Ueda, H., Hanai, K. and Naito, T. *The Improvements of Fatigue Durability of Carburised Steels with Surface Structure Anomalies by Shot Peening: Advances in Surface Treatments, Technology, Applications, Effects*, Vol. 2, A. Niku-Lari, ed., Pergamon Press, pp. 69–74, 1986.

13 Solod, G. I. et al. "Improving Bending Fatigue of Large-Module Gear Teeth," *Russian Engineering Journal*, LII(1):19–22, 1972.

14 Wilks, T. P. Ph.D. Thesis, Newcastle University, UK, 1985.

Index

For Product Safety Concerns and Information please contact our EU
representative GPSR@taylorandfrancis.com Taylor & Francis Verlag GmbH,
Kaufingerstraße 24, 80331 München, Germany

Printed and bound by CPI Group (UK) Ltd, Croydon, CR0 4YY

01/05/2025

01858591-0001